D1521317

SWITCHED-CURRENT DESIGN AND IMPLEMENTATION OF OVERSAMPLING A/D CONVERTERS

THE KLUWER INTERNATIONAL SERIES IN ENGINEERING AND COMPUTER SCIENCE

ANALOG CIRCUITS AND SIGNAL PROCESSING

Consulting Editor

Mohammed Ismail

Ohio State University

Related Titles:

DESIGN OF LOW-VOLTAGE, LOW-POWER OPERATIONAL AMPLIFIER CELLS, *Ron Hogervorst, Johan H. Huijsing*, ISBN: 0-7923-9781-9

VLSI-COMPATIBLE IMPLEMENTATIONS FOR ARTIFICIAL NEURAL NETWORKS, *Sied Mehdi Fakhraie, Kenneth Carless Smith*, ISBN: 0-7923-9825-4

CHARACTERIZATION METHODS FOR SUBMICRON MOSFETs, edited by *Hisham Haddara*, ISBN: 0-7923-9695-2

LOW-VOLTAGE LOW-POWER ANALOG INTEGRATED CIRCUITS, edited by *Wouter Serdijn*, ISBN: 0-7923-9608-1

INTEGRATED VIDEO-FREQUENCY CONTINUOUS-TIME FILTERS: *High-Performance Realizations in BiCMOS*, *Scott D. Willingham, Ken Martin*, ISBN: 0-7923-9595-6

FEED-FORWARD NEURAL NETWORKS: *Vector Decomposition Analysis, Modelling and Analog Implementation*, *Anne-Johan Annema*, ISBN: 0-7923-9567-0

FREQUENCY COMPENSATION TECHNIQUES LOW-POWER OPERATIONAL AMPLIFIERS, *Ruud Easchauzier, Johan Huijsing*, ISBN: 0-7923-9565-4

ANALOG SIGNAL GENERATION FOR BIST OF MIXED-SIGNAL INTEGRATED CIRCUITS, *Gordon W. Roberts, Albert K. Lu*, ISBN: 0-7923-9564-6

INTEGRATED FIBER-OPTIC RECEIVERS, *Aaron Buchwald, Kenneth W. Martin*, ISBN: 0-7923-9549-2

MODELING WITH AN ANALOG HARDWARE DESCRIPTION LANGUAGE, *H. Alan Mantooth, Mike Fiegenbaum*, ISBN: 0-7923-9516-6

LOW-VOLTAGE CMOS OPERATIONAL AMPLIFIERS: *Theory, Design and Implementation*, *Satoshi Sakurai, Mohammed Ismail*, ISBN: 0-7923-9507-7

ANALYSIS AND SYNTHESIS OF MOS TRANSLINEAR CIRCUITS, *Remco J. Wiegerink*, ISBN: 0-7923-9390-2

COMPUTER-AIDED DESIGN OF ANALOG CIRCUITS AND SYSTEMS, *L. Richard Carley*, Ronald S. Gyurcsik, ISBN: 0-7923-9351-1

HIGH-PERFORMANCE CMOS CONTINUOUS-TIME FILTERS, *José Silva-Martínez, Michiel Steyaert, Willy Sansen*, ISBN: 0-7923-9339-2

SYMBOLIC ANALYSIS OF ANALOG CIRCUITS: *Techniques and Applications*, *Lawrence P. Huelsman, Georges G. E. Gielen*, ISBN: 0-7923-9324-4

DESIGN OF LOW-VOLTAGE BIPOLAR OPERATIONAL AMPLIFIERS, *M. Jeroen Fonderie, Johan H. Huijsing*, ISBN: 0-7923-9317-1

STATISTICAL MODELING FOR COMPUTER-AIDED DESIGN OF MOS VLSI CIRCUITS, *Christopher Michael, Mohammed Ismail*, ISBN: 0-7923-9299-X

SELECTIVE LINEAR-PHASE SWITCHED-CAPACITOR AND DIGITAL FILTERS, *Hussein Baher*, ISBN: 0-7923-9298-1

ANALOG CMOS FILTERS FOR VERY HIGH FREQUENCIES, *Bram Nauta*, ISBN: 0-7923-9272-8

ANALOG VLSI NEURAL NETWORKS, *Yoshiyasu Takefuji*, ISBN: 0-7923-9273-6

ANALOG VLSI IMPLEMENTATION OF NEURAL NETWORKS, *Carver A. Mead, Mohammed Ismail*, ISBN: 0-7923-9049-7

AN INTRODUCTION TO ANALOG VLSI DESIGN AUTOMATION, *Mohammed Ismail, José Franca*, ISBN: 0-7923-9071-7

SWITCHED-CURRENT DESIGN AND IMPLEMENTATION OF OVERSAMPLING A/D CONVERTERS

by

Nianxiong Tan
Microelectronics Research Center
Ericsson Components AB
Sweden

KLUWER ACADEMIC PUBLISHERS
Boston / Dordrecht / London

Distributors for North America:
Kluwer Academic Publishers
101 Philip Drive
Assinippi Park
Norwell, Massachusetts 02061 USA

Distributors for all other countries:
Kluwer Academic Publishers Group
Distribution Centre
Post Office Box 322
3300 AH Dordrecht, THE NETHERLANDS

Library of Congress Cataloging-in-Publication Data

A C.I.P. Catalogue record for this book is available
from the Library of Congress.

Printed on acid-free paper.

Printed in the United States of America

Contents

Preface

"A book of practical designs,
by a designer,
and for designers."

Designing analog circuits in a digital CMOS process is of great interest for it provides the possibility of integrating a whole signal processing system into a single chip. The switched-current (SI) technique is an analog sampled-data technique that fully exploits the digital CMOS process. Although in a digital CMOS process it is possible to use the traditional switched-capacitor (SC) technique with the aid of linear capacitors created by wire parasitic capacitance, the SI technique is a viable alternative to the traditional SC technique with both pros and cons. The major advantages of the SI technique are the high-speed and low-supply-voltage operations, while the major disadvantages are the relatively larger noise and less accuracy compared with the SC technique. In oversampling A/D converters, the advantages of the SI technique can be fully utilized while the disadvantages can be suppressed to a certain extent.

The purpose of the book is to present the design and implementation of oversampling A/D converters in the digital CMOS process using the SI technique. The emphases are given to the comparison of the SC and SI design, the practical aspects of SI circuits and systems, and the system design of oversampling A/D converters for the SI implementation. Also presented are the measurement results of many test chips, demonstrating high-speed (over 100 MHz) and ultra-low-voltage (1.2 V) operations of SI circuits and systems.

This book especially addresses readers who are interested in analog circuit design. For those who know about the SI technique, the book can be served as a practical guidance in that many circuit implementations are included rather than tedious mathematical derivations or theoretical arguments; for those who are familiar with the SC technique, the book can be used to study the SI technique in that in the book the comparison of the SC and SI techniques is highlighted. This book can be used as a reference book for the industrial analog design and for the advanced graduate course on the current-mode techniques. This is a book of practical designs, by a designer, and for designers.

The book consists of eight chapters.

In *Chapter I* we introduce the principle of the SI technique. We compare the SC and SI technique concerning the process cost, the speed, the requirement on the supply voltage, and the dynamic range, etc. We draw special attentions to this comparison when the SC technique is used in a digital CMOS process with the aid of linear capacitors created by wire parasitic capacitance (rather than by using more expensive double-poly or poly/N+ linear capacitors). The influence of the CMOS scaling on the SC and SI techniques is also highlighted.

In *Chapter II* we concentrate on the nonidealities in the basic SI circuits and compare them with nonidealities in the basic SC circuits. The mismatch errors, finite input-output conductance ratio errors, settling errors, clock feedthrough errors, drain-gate parasitic capacitive coupling errors, and noise errors are outlined and compared with the SC counterparts.

In *Chapter III* we outline techniques to increase the performance of basic SI circuits and present 6 measured SI circuits including first-generation SI circuits with clock feedthrough compensated, low-voltage fully differential SI circuits, low-voltage class-AB fully differential SI circuits, low-voltage high-speed SI circuits, ultra low-voltage (1.2-V) SI circuits, and low-voltage high-speed two-step SI circuits. The common-mode feedforward (CMFF) technique is employed in some of the fully differential SI circuits, and the low-power CMFF technique is introduced. We also touch upon the SI technique in a digital BiCMOS process. Other existing techniques to increase the performance of SI circuits are discussed and compared briefly as well.

In *Chapter IV* the system design of SI oversampling A/D converters (or oversampling delta-sigma modulators) is addressed. We explain why the SI technique is especially suitable for oversampling A/D converter design. By emphasizing the similarity of and difference between the SC and SI implementations, we present the technique of modifying an SC oversampling delta-sigma modulator architecture to an SI oversampling delta-sigma modulator architecture. We also present 4 different oversampling delta-sigma modulator architectures tailored for the SI implementation including a second-order, two-stage fourth-order, single-stage fourth-order, and system-level chopper-stabilized delta-sigma modulators. To increase the dynamic range of SI delta-sigma modulators in which the thermal noise at the input rather than the quantization noise limits the dynamic range, we present a current scaling technique which increases the dynamic range without a drastic increase in the power consumption and chip area. The power scaling technique can be applied to any SI delta-sigma modulator architectures.

In *Chapter V* we discuss the building blocks for SI oversampling A/D converters including SI integrators, SI differentiators, current quantizers, 1-bit D/A converters, and clock generators. To facilitate the measurement, on-chip voltage-to-current (V/I) converters are needed. We also present the technique of using the on-chip V/I converter as a low pass filter and a continuous-time filtering technique for current-mode A/D converters. To reduce the high-frequency noise generated in an oversampling A/D converter (or delta-sigma modulator), digital decimation filters are needed. The design of high-speed and low-power digital decimation filters is also covered in this chapter. To make a fully functional ultra low-voltage SI oversampling A/D converters, we need some other auxiliary circuits such as clock voltage doublers, high voltage generators, and level-shifting circuits, etc. These building blocks are touched upon in this chapter as well.

To make functioning SI circuits, we need to consider practical aspects. *Chapter VI* is devoted to this purpose. Preceded by the simulation setup consideration, the clocking strategy, loading effects, and resetting for SI circuits are presented. Also detailed in this chapter are the basic analog layout techniques with an emphasis on the matching and the mixed analog-digital floor planning and layout with an emphasis on the noise coupling.

In *Chapter VII* we present the measurement results of nine test chips of SI oversampling delta-sigma modulators and compare the measurement results with theoretical expectations. We also compare the performance of SI delta-sigma modulators with the state-of-the-art SC delta-sigma modulators, highlighting the advantages of the SI implementations such as higher bandwidths and lower supply voltages.

Chapter VIII concludes the book.

I would like to thank all the people who helped me in making the book possible. They are the late Dr. Sven Eriksson, Prof. Lars Wanhammar, Håkan Träff, Mikael Gustavsson, Jacob Wikner, and Anders Ihlström of Linköping University, Linköping, Prof. Hannu Tenhunen of the Royal Institute of Technology, Stockholm, Giti Amozandeh, Allan Olson, and Helge Stenström of Ericsson Components AB, Stockholm, Bengt Jonsson of Ericsson Radio Systems AB, Stockholm, and Yonghong Gao of Tsinghua University, Beijing.

SWITCHED-CURRENT DESIGN AND IMPLEMENTATION OF OVERSAMPLING A/D CONVERTERS

Chapter I: Basics of the SI Technique

1.1. PRINCIPLE OF THE SI TECHNIQUE

The development of VLSI process technologies has had a great impact on the design philosophy of integrated circuits and systems. Traditionally, analog circuits had a lion's share in signal processing, but recent years have witnessed the steady replacement of digital signal processing (DSP) circuits for analog signal processing (ASP) circuits due to the advance of process technologies and advantages of digital circuits. However, the real world is of analog form and therefore, analog circuits will always exist and play an indispensable role in the information society.

There exist different techniques in designing analog circuits. The most recent advance is the switched-current (SI) technique [1]. The SI technique was first introduced in the late eighties [2, 3]. Let's start by looking at the principle of the SI technique.

For an MOS transistor in its saturation region, we have the following relationship [4]

$$I_D = \frac{\mu C_{ox}}{2} \cdot \frac{W}{L} \cdot (V_{GS} - V_T)^2 \cdot (1 + \lambda V_{DS}), \; 0 < (V_{GS} - V_T) \leq V_{DS} \qquad (1.1)$$

where I_D is the drain current, μ is the charge mobility, C_{ox} is the gate capacitance per square, W is the transistor width, L is the transistor length, V_{GS} is the gate source voltage, V_T is the threshold voltage, λ is the channel length modulation factor, and V_{DS} is the drain source voltage.

If we neglect the channel length modulation factor λ, we have the following simple equation describing the relationship between the gate source voltage and the drain current.

$$I_D = \frac{\mu C_{ox}}{2} \cdot \frac{W}{L} \cdot (V_{GS} - V_T)^2, \; 0 < (V_{GS} - V_T) \leq V_{DS} \qquad (1.2)$$

It is seen from equation (1.2) that the drain current of an MOS transistor is solely determined (to the first order) by the gate source voltage being applied to it. Since the impedance at the gate of an MOS transistor is very large, a voltage can be memorized and therefore the drain current can be memorized. This is the starting point of any SI circuits.

In figure 1.1 we show the basic building blocks in SI circuits, i.e., the SI memory cells. Shown in figure 1.1 (a) is the first generation SI memory cell [3] with controlling clock signals, and shown in figure 1.1 (b) is the second generation SI memory cell [2] with controlling clock signals.

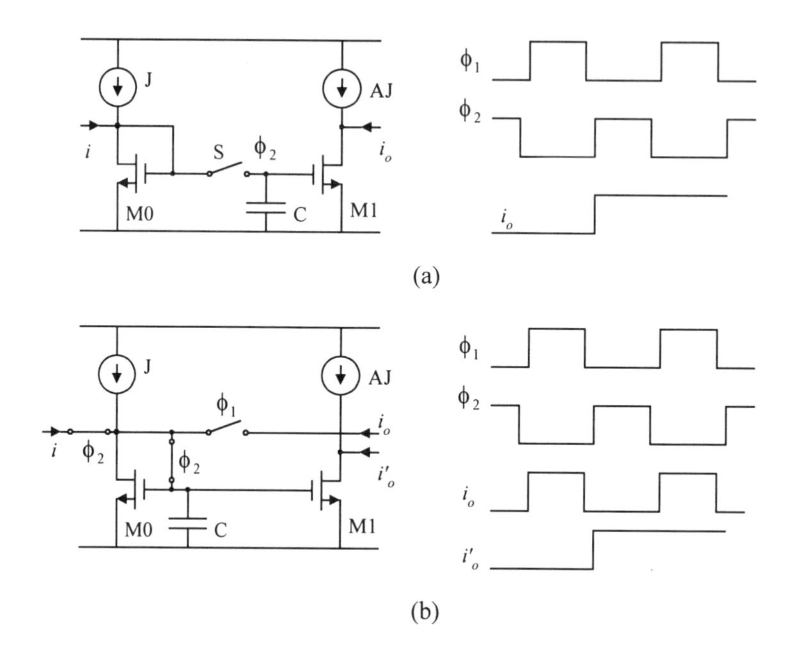

Fig. 1.1. Switched-current memory cells.
(a) first-generation and (b) second-generation.

Referring to figure 1.1 (a), if we do not have the switch S controlled by clock phase Φ_2, the first-generation SI memory cell would function as a current mirror. If the bias current for transistor M1 is *A* times larger than the bias current for transistor M0 and the dimension ratio of M1 is A times larger than that of M0, the output current i_o is A times larger than the input current i. With the switch added, a track and hold function is realized.

During clock phase Φ_2 when the switch S is on, the first-generation SI memory cell shown in figure 1.1 (a) behaves as a current mirror with an amplification factor A. During clock phase Φ_1 when the switch S is off, the gate of transistor M1 is isolated. Since the impedance at an MOS transistor is very high, the gate voltage of the MOS transistor M1 is 'frozen' at the value at the moment when the switch S opens, i.e., the gate voltage is held by the parasitic capacitance at the gate of the MOS transistor M1. To the first order, the drain current is determined by the gate voltage according to equation (1.2). The drain current of M1 and therefore the output current i_o are the corresponding values at the moment when the switch S opens. Thereby, a track-and-hold function is realized by the first-generation SI memory cell. Shown in figure 1.2 (a) is the wave form of the input and output currents. The input current is a sinusoid and the amplification factor A is assumed to be unity. Notice that the direction of the output current is inverted just as in a current mirror.

Shown in figure 1.1 (b) is the second-generation SI memory cell. The only difference is the use of a single transistor M0 both as the input and as the output transistor.

On clock phase Φ_2, transistor M0 is diode connected and conducts a current $J + i$ (where J is the bias current and i is the input current). On clock phase Φ_1, the gate is isolated and the gate voltage is held by the gate parasitic capacitor C and therefore, the drain current $J + i$ sustains and the output current $i_o = i$. As transistor M0 is used alternatively as the input and output device, i_o is available only during clock phase Φ_2. The wave form of the second-generation SI memory cell is shown in figure 1.2 (b). The input current is a sinusoid. Notice that the direction of the output current is inverted just as in a current mirror.

A current mirror can be used to provide a current for a whole period, as shown in figure 1.1 (b). A scaling factor can also be included in the output current mirror.

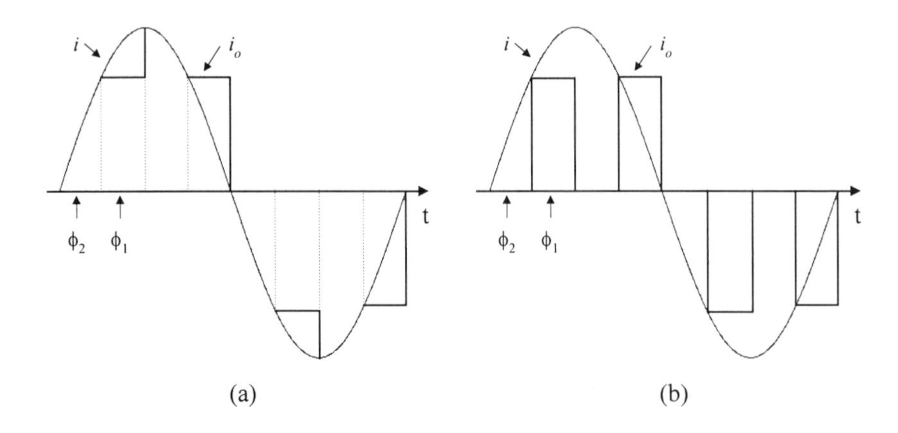

Fig. 1.2. Wave forms of switched-current memory cells. (a) first-generation and (b) second-generation.

1.2. Comparison of the SI And SC Techniques

A. Process cost

In SI circuits as in the SI memory cells, the function of capacitors is to hold the gate voltage temporarily and therefore, parasitic gate capacitors suffice and linear capacitors are not required.

For the switched-capacitor (SC) circuits [5], the processing of signals relies on the charge transfer and the voltage excursion. The linearity of the capacitors directly determines the linearity of the SC circuits. Therefore, linear capacitors are usually needed for SC circuits, though there are efforts in using nonlinear MOS gate capacitors in SC circuits [6-8]. The use of nonlinear MOS gate capacitors in SC circuits calls for linearization of the nonlinear MOS gate capacitors or cancellation of nonlinear components in the MOS gate capacitors. So far there have not been reported very good measurement results.

To create linear capacitors, double poly layers are usually needed. But in a 'standard' digital CMOS process, the second poly layer is not available. Therefore, extra masks are needed to create linear capacitors in the standard digital CMOS process and this increases the cost significantly not only due to the extra process steps but also due to the reduced yield. (Notice it is also possible to create linear metal-to-metal capacitors by thinning the oxide in-between, or create linear poly-to-N+ (or P+) capacitors by heavily doping the active areas beneath the poly gates. All these kinds of linear capacitors call for extra process steps, though their properties differ.)

It is also possible to design SC circuits in a standard digital CMOS process with the aid of linear capacitor formed by poly and metal plates. However, the linearity is not as good as that of double-poly capacitors and the capacitance per square of this kind of linear capacitors is low. (The sheet capacitance can be increased by sandwiching more conducting plates if more metal layers are available in the process.) It is area-consuming to create this kind of linear capacitors. The matching of this kind of capacitors is poor as well, even compared with the matching of MOS transistors [9].

As the trend of CMOS scaling indicates, the gate capacitance per square is getting bigger and bigger (to increase the driving capability of MOS transistors) while the wire parasitic capacitance per square is getting smaller and smaller (to increase the speed and reduce the power dissipation). The use of the capacitors formed by wires may not be a good choice, because they are too area-consuming compared with the MOS gate capacitors.

B. High speed operation

In SI circuits, the signal carrier is the current. Therefore the impedance at every node is low. Due to the inherent low impedance, SI circuits are very suitable for high frequency applications. The settling time of SI memory cells shown in figure 1.1 is determined by the time constant given by

$$\tau = \frac{C}{g_{m0}} \tag{1.3}$$

where C is the total capacitance at the gate of the memory transistor M0 and g_{m0} is the transconductance of the memory transistor M0.

If the width of transistor M1 is A times larger than that of transistor M0 and the length is the same, and the total parasitic capacitance C at the gate of transistor M0 is dominated by the gate capacitance of transistors M0 and M1, the total parasitic capacitance C is given [4] by (assume that both transistors are in the saturation region)

$$C = \frac{2}{3} \cdot C_{ox} \cdot (1 + A) \cdot W \cdot L \tag{1.4}$$

where C_{ox} is the gate capacitance per square, A is the width ratio of transistor M1 and M0, W is the width of transistor M0, and L is the length of transistors M0 and M1.

The transconductance of transistor M0 in the saturation region is given [4] by

$$g_{m0} = \mu C_{ox} \cdot \frac{W}{L} \cdot (V_{GS} - V_T) \tag{1.5}$$

where μ is the charge mobility, C_{ox} is the gate capacitance per square, W is the width of transistor M0, L is the length of transistor M0, V_{GS} is the gate source voltage, and V_T is the threshold voltage.

By combining equations (1.3) ~ (1.5), we have the settling time constant given by

$$\tau = \frac{2}{3} \cdot \frac{(1 + A) \cdot L^2}{\mu \cdot (V_{GS} - V_T)} \tag{1.6}$$

Based on the parameters of Ericsson's in-house 0.6 μm CMOS process, the time constant is around 64 ps for $A = 1$, $L = 0.6$ μ and the gate source voltage overhead $V_{GS} - V_T = 0.5\,\text{V}$. This indicates the potential for high speed applications of the SI technique. Parasitic and nonlinearities in the basic memory cells make it difficult to operate SI circuits in GHz range. However, it is achievable to operate SI circuits at a clock frequency close or even over 100 MHz [10].

For SC circuits, operational amplifiers are usually needed and they dictate the speed of SC circuits. With a high DC gain requirement as for most SC circuits, it is very difficult to push the unity-gain bandwidth of operational amplifiers to a very high frequency and at the same time to have a linear settling and have a large phase margin to guarantee stability.

C. Low supply voltage operation

In SI circuits, the signal carriers are current samples and the supply voltage does not limit the signal range. Therefore, SI circuits are suitable for low-voltage operations.

Referring to figure 1.1, the requirement of the supply voltage is given by

$$V_{dd} \geq V_T + \left(V_{GS} - V_T\right) \cdot \sqrt{1 + m_i} + \left(V_{GS} - V_T\right)_J \tag{1.7}$$

where V_T is the threshold voltage of the memory transistor M0, $(V_{GS} - V_T)$ is the quiescent saturation voltage of M0, m_i is the input modulation index (the highest input current vs. the bias current J), and $(V_{GS} - V_T)_J$ is the quiescent saturation voltage of the transistor forming the bias current source J. With equation (1.7) satisfied, the memory transistor M0 is guaranteed to be in its saturation region with different input currents.

From equation (1.7) it is seen that the supply voltage is only required to be larger than one threshold voltage. It is proved that it is feasible to operate SI circuits with a supply voltage less than twice the threshold voltage [11].

For most SC circuits, there is usually a need for operational amplifiers. It is very difficult to design high gain high speed operational amplifiers when the supply voltage is reduced close to twice the threshold voltage.

D. Noise, dynamic range, and power consumption

For MOS transistors, there are two major types of noise, flicker noise (or 1/*f* noise) and thermal noise [4]. The low frequency noise, flicker noise can be reduced by using the correlated double sampling or chopper-stabilization techniques [4, 5]. For wide signal band applications, flicker noise usually does not dominate. It is thermal noise that overwhelms.

When we sample a signal, all the thermal noise folds into the signal band. It imposes a fundamental limitation in SC circuits. When sampling a voltage signal on a capacitor C, the total noise power is given by

$$\overline{v_{rms}^2} = \frac{kT}{C} \tag{1.8}$$

where k is the Boltzmann constant, T is the absolute temperature, and C is the sampling capacitance. Notice the total noise power is independent of the switch-on resistance and the sampling frequency.

For SI circuits, we have a similar fundamental limitation.

Neglecting the contribution of low frequency noise such as flicker noise, we only consider thermal noise. The current noise power spectral density of an MOS transistor in its saturation region can be approximated [5] by

$$\frac{\overline{i_n^2}}{\Delta f} = \frac{8}{3} \cdot k \cdot T \cdot g_m \tag{1.9}$$

where k is the Boltzmann constant, T is the absolute temperature, g_m is the transconductance of the MOS transistor.

The noise contribution to the memory cells shown in figure 1.1 is both from the memory transistor itself and from the current source, neglecting the contribution from the load. Therefore the current noise density of the memory cells shown in figure 1.1 is given by

$$\frac{\overline{i_n^2}}{\Delta f} = \frac{8}{3} \cdot k \cdot T \cdot \left(g_{m0} + g_{mJ}\right) \tag{1.10}$$

where k is the Boltzmann constant, T is the absolute temperature, g_{m0} is the transconductance of the memory transistor M0, and g_{mJ} is the transconductance of the transistor forming the current source J.

The noise bandwidth of a single pole system such as the memory cells shown in figure 1.1 is given by

$$BW_n = \frac{\pi}{2} \cdot f_{pole} = \frac{\pi}{2} \cdot \frac{1}{2\pi\tau} = \frac{g_{m0}}{4C} \tag{1.11}$$

where f_{pole} is the pole frequency, τ is the time constant, g_{m0} is the transconductance of the memory transistor M0, and C is the total capacitance at the gate of the transistor M0.

Therefore, the total current noise power is the product of the noise power spectral density and the noise bandwidth, i.e.,

$$\overline{i_{rms}^2} = \frac{\overline{i_n^2}}{\Delta f} \cdot BW_n = \frac{kT}{C} \cdot \left\{ \frac{2}{3} g_{m0}^2 \cdot \left[1 + \frac{g_{mJ}}{g_{m0}} \right] \right\} \tag{1.12}$$

If we refer the current noise to the gate of the transistor M0, we have the total voltage noise power

$$\overline{v_{rms}^2} = \frac{\overline{i_{rms}^2}}{g_{m0}^2} = \frac{kT}{C} \cdot \left\{ \frac{2}{3} \cdot \left[1 + \frac{g_{mJ}}{g_{m0}} \right] \right\} \approx \frac{kT}{C} \tag{1.13}$$

The approximation takes into consideration that the transconductance of the current source transistor is usually smaller or in the same range as the transconductance of the memory transistor M0.

When we operate the SI memory cells with a sampling clock, the sampling operation does not change the total noise power but only redistributes it. If we neglect the noise contribution from the switch-on resistance of the switch transistors [1], the total noise power is in the same range as given by equation (1.13).

Notice that the fundamental limitation of thermal noise is the same for both SC (equation (1.8)) and SI circuits (equation (1.13)). Usually, the parasitic capacitor in SI circuits is considerably smaller than the sampling capacitor in SC circuits, and therefore, SI circuits are more noisy than SC circuits.

It is possible to use a large capacitance created by the gate of MOS transistors in SI circuits at the cost of speed or power consumption in such a way that the noise values determined by equations (1.8) and (1.13) are in the same range. However, the dynamic range of SI circuits is still considerably lower than that of SC circuits. The reason is as follows.

If the SC and SI circuits have the same voltage noise power, the dynamic range is then determined by the input signal (voltage) swing. For SC circuits, the signal swing is in the same range as the supply voltage, but in SI circuits, the voltage swing referred to the gate from the signal (current) swing is considerably smaller than the supply voltage. (Notice that we refer both the input current and the current noise to the gate of the memory transistor. It is equivalent to introducing a scaling factor, g_{m0} in the dynamic range calculation.) Therefore the dynamic range of SI circuits is usually smaller than that of SC circuits.

Referring to equation (1.8), the only way to increase the dynamic range of SC circuits (where the thermal noise is the fundamental limitation) is to increase the sampling capacitance. For every doubling of the sampling capacitance, the thermal noise power decreases by 3 dB and it is only possible to increase the dynamic range by 3 dB. To drive the increased capacitance, more power is needed for the operational amplifiers. The real problem is not just the power consumption, but the feasibility of maintaining the speed with an increased capacitive load for the operational amplifiers. The problem deteriorates when the supply voltage is pressed down.

For SI circuits, it is possible to increase the dynamic range while maintaining the speed at the cost of power consumption.

Suppose we increase the width of all the transistors by a factor of α while keeping the transistor length and at the same time we increase the bias current by a factor of α as well. The capacitance will increase by a factor of α and therefore the voltage noise power referred at the gate will decrease by a factor of α according to equation (1.13).

Referring to equation (1.2), we see that the gate source voltage will not change since we increase both the drain current and the transistor width. Therefore, the voltage swing capability at the gate of the memory transistor M0 will maintain. The total effect is that for every doubling of the bias current (by doubling the transistor width to maintain the gate source voltage), the dynamic range increases by 3 dB due to the decrease in the thermal noise according to equation (1.13) and the same voltage swing capability at the gate (i.e., the same signal handling capability).

Unlike in SC circuits, however, the speed will not suffer. The transconductance g_{m0} will increase by a factor of α according to equation (1.5). Therefore the settling time constant will not change according to equation (1.3) since both the capacitance C and transconductance g_{m0} increase by a factor of α.

Alternatively, we can think of the current swing and the current noise in SI circuits (instead of referring them to the gate) and derive the same conclusion.

The largest input is limited by the bias current while the smallest input is determined by the noise. Here, we suggest a 'current scaling' principle. Referring to figure 1.1, if we increase the bias current by α times and the width of the memory transistor by α times, all the nodal voltages remain unchanged according to equation (1.2). Therefore, the largest input current is increased by α times. Another very important issue is the speed. Since we increase the transistor width by α times, the gate capacitance C increases by α times according to equation (1.4) but at the same time, the transconductance g_{m0} increases by α times as well according to equation (1.5). Therefore, the time constant remains unchanged according to equation (1.3) and the speed does not degrade. Now, let's look at the smallest input current when we increase the bias current.

If the noise generated by the SI circuits is not the dominant factor of limiting the lowest input, every doubling of the bias current enables the largest input to increase by 2 times and therefore, the dynamic range increases by 6 dB. This is usually the case when we integrate analog and digital circuits on the same chip. The switching noise from the neighboring digital circuits may dominate.

If the noise generated by the SI circuits is the dominant factor, we have a different story. First of all, the $1/f$ noise is usually of no concern in the second-generation SI circuits due to the correlated double sampling [1] and the $1/f$ noise in the first-generation SI circuits can also be reduced by using the chopper stabilization technique. For wideband applications, the influence of the $1/f$ noise is minor. We only concern with thermal noise here.

The noise bandwidth of the SI memory cells is given by equation (1.11), which is constant while we apply the current scaling. However, the current noise power spectral density given by equation (1.10) increases by a factor of α. Therefore, for every doubling of the bias current (the transconductance increases by 2 times), the noise power increases by 3 dB, while the largest input increases by 6 dB. The total effect is that the dynamic range increases by 3 dB for every doubling of the bias current, if the thermal noise of SI circuits determines the smallest input.

This current scaling principle can be utilized at the system level to increase the dynamic range with a moderate increase in power consumption if we can treat different parts separately [12-14].

E. Accuracy

The second-generation SI memory cell itself does not have the mismatch problem since the same memory transistor is used alternatively as the input and output device. But general SI circuits, either based on the first-generation or on the second-generation memory cells have the mismatch problem, because current mirrors are always needed. In the CMOS process optimized for high density high speed digital circuits, it is usually difficult to make the transistor matching better than 0.1% even with large transistors. On the contrary, the matching of linear capacitors can be made close to 0.01%. Therefore, the SI technique is not very suitable for high performance filtering applications in that the error in coefficients is large, though some SI filters can be found in the literature [1].

F. Complexity, modularity and testability

In SC circuits, operational amplifiers are needed which are relatively complex. In SI circuits, we do not need operational amplifiers which makes the complexity of SI circuits relatively low.

Another nice feature of the SI technique is the modularity. Different loads only need different output current mirrors and there is usually no need to change the circuit configuration. If the load is too large, it only reduces the speed. However, for SC circuits, different loads may require different architectures for the operational amplifiers because for most high speed SC circuits, single-stage operational amplifiers are used and the load is usually used to determined the frequency response. Large variation in load capacitance may cause instability or even oscillation in SC circuits, not just degrades the speed.

SI circuits can in principle be tested by only sensing the branch currents and supply currents. Built-in self tests can also be done [15]. For SC circuits, we usually need to access nodal voltages to test the circuits. It is relatively difficult.

1.3. SUMMARY

The SI technique is a viable technique to design analog sampled-data circuits in a digital CMOS process. It has both pros and cons compared with the SC technique. The major advantages of the SI technique are low cost, high speed, possibility of very low voltage operation, and modularity. The major disadvantages of the SI technique are the less accuracy, noisiness, and lower dynamic range. In oversampling A/D converters, the advantages can be fully exploited and the disadvantages can be suppressed to a certain extent [16, 17].

Chapter II: Nonidealities in SI Circuits

2.1. Introduction

The principle of the SI technique has been described in Chapter I. We have also touched upon certain nonidealities of the SI technique in order to compare the SI technique with the SC technique, highlighting the pros and cons of the SI technique. In this chapter, we will detail the nonidealities of SI circuits including the mismatch errors, finite input-output conductance ratio errors, settling errors, clock feedthrough errors, drain-gate parasitic capacitve coupling errors, and noise errors. Comparisons with the nonideal behaviors of SC circuits will be given when applicable as well.

2.2. Mismatch Errors

Though the second-generation SI memory cell shown in figure 1.1 (b) does not suffer from the mismatch problem in that the same transistor is used as both the input and output device, general SI circuits suffer from the mismatch problem. Any coefficients needed in an SI system are realized by current mirrors. Therefore, the matching property of SI circuits is basically the same as that of current mirrors.

For an MOS transistor in its saturation region, we have the following square-law relationship [4]

$$\begin{aligned} I_D &= \frac{\mu C_{ox}}{2} \cdot \frac{W}{L} \cdot \left(V_{GS} - V_T\right)^2 \cdot \left(1 + \lambda V_{DS}\right),\ 0 < \left(V_{GS} - V_T\right) \le V_{DS} \\ &= \frac{\beta}{2} \cdot \left(V_{GS} - V_T\right)^2 \cdot \left(1 + \lambda V_{DS}\right),\ 0 < \left(V_{GS} - V_T\right) \le V_{DS} \end{aligned} \tag{2.1}$$

where I_D is the drain current, μ is the charge mobility, C_{ox} is the gate capacitance per square, W is the transistor width, L is the transistor length, V_{GS} is the gate source voltage, V_T is the threshold voltage, λ is the channel length modulation factor, V_{DS} is the drain source voltage, and β is the transconductance parameter. The transconductance parameter β is given by

$$\beta = \mu C_{ox} \cdot \frac{W}{L} \tag{2.2}$$

Suppose we are required to duplicate the current as shown in figure 2.1, where transistors M0 and M1 are supposed to be matched. The load for transistor M1 is also shown in Figure 2.1.

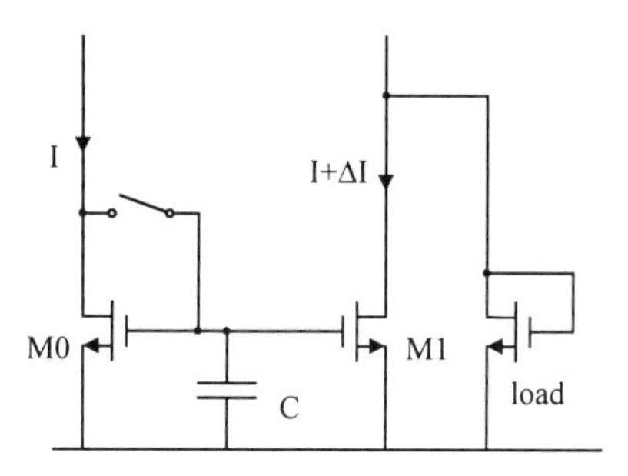

Fig. 2.1. Second-generation SI memory cell with a duplicated output current.

Due to the local variation in all the parameters given by equations (2.1) and (2.2), there is a current variation ΔI. Suppose all the variations are small and are designated with a Δ before the parameters. Then the relative variation in the current is given by

$$\begin{aligned} \frac{\Delta I}{I} &= \frac{\Delta\beta}{\beta} + \frac{2\Delta V_{GS}}{V_{GS}-V_T} - \frac{2\Delta V_T}{V_{GS}-V_T} + \frac{\Delta\lambda}{\frac{1}{V_{DS}}+\lambda} + \frac{\Delta V_{DS}}{V_{DS}+\frac{1}{\lambda}} \\ &= \frac{\Delta\mu}{\mu} + \frac{\Delta C_{ox}}{C_{ox}} + \frac{\Delta\frac{W}{L}}{\frac{W}{L}} + \frac{2\Delta V_{GS}}{V_{GS}-V_T} - \frac{2\Delta V_T}{V_{GS}-V_T} + \frac{\Delta\lambda}{\frac{1}{V_{DS}}+\lambda} + \frac{\Delta V_{DS}}{V_{DS}+\frac{1}{\lambda}} \end{aligned} \tag{2.3}$$

Since the two transistors M0 and M1 have the same nominal size and same nominal terminal potentials, therefore, the difference in the charge mobility is small. Since the two transistor M0 and M1 have the same nominal channel length, the difference in the channel length modulation is small. By simplifying equation (2.3), we have

$$\frac{\Delta I}{I}=\frac{\Delta C_{ox}}{C_{ox}}+\frac{\Delta\frac{W}{L}}{\frac{W}{L}}-\frac{2\Delta V_T}{V_{GS}-V_T}+\frac{2\Delta V_{GS}}{V_{GS}-V_T}+\frac{\Delta V_{DS}}{V_{DS}+\frac{1}{\lambda}} \tag{2.4}$$

The first three terms are determined by the given process, and the last two terms are layout-related.

The first term is due to the difference in the oxide thickness at different locations, the second term is due to the uncertainty of the device geometry, and the third term is due to the spread of the threshold voltage. With a lower supply voltage, the spread of the threshold voltage dominates the mismatch because V_{GS} - V_T is very small. Notice that due to the random nature of these uncertainties, their influence can not be counteracted even though the sign of the third term is negative.

The difference in the gate source voltages of transistors M0 and M1 is due to the parasitic resistance in an actual layout. Though the gates of transistors M0 and M1 do not carry any static currents, the source currents must be supplied by the supply via wires. The wire connecting the sources of the transistors M0 and M1 has parasitic resistance which inflicts the difference in the gate source voltages. The influence is severe when the supply voltage is pressed down, and/or several transistors are needed to be matched (therefore the distance is increased). Suppose V_{GS} - V_T = 0.2 V, a mismatch of 0.5 mV in the drain source voltage destroys the possibility of achieving a 8-bit matching even with the assumption that everything else was perfectly matched. If a 0.5 mA current flows between the sources of transistors M0 and M1, only a 1-ohm parasitic resistance suffices to destroy the possibility of achieving the 8-bit matching. One ohm resistance in a CMOS process usually means only one contact or a 20-μm long wire with 1-μm width. It is therefore extremely important to emphasize the layout. Without a careful floor planning and layout, the matching of analog circuits is poor.

The difference in the drain source voltages is due to the difference in the source and drain potentials. The difference in the source potentials is due to the parasitic resistance, and the difference in the drain potentials is due to the fact that the load does not have a zero input impedance. Their influence on matching is significant due to the large channel length modulation factor λ when short channel devices are used to boost the speed. Cascoding reduces the influence significantly.

In SC circuits, the accuracy is usually determined by the matching of capacitors. The matching of capacitors is determined by

$$\frac{\Delta C}{C} = \frac{\Delta C_{ox}}{C_{ox}} + \frac{\Delta A}{A} \tag{2.5}$$

where C is the nominal capacitance, C_{ox} is the capacitance per square, A is the nominal capacitance area, and Δ indicates the local variation of the parameters.

For double-ploy capacitors, the variation of C_{ox} is comparable to the variation of the MOS gate capacitance, and the area uncertainty is also comparable. Due to the extra terms in equations (2.4) (the variation in the threshold voltage V_T), the matching of transistors is usually poorer than the matching of linear capacitors.

For linear capacitors formed by the parasitic capacitance between wires, the variation in C_{ox} is considerably larger than the variation of the MOS gate capacitance and the matching of the geometry is also poorer than the matching of the MOS transistor geometry due to the roughness of metal surfaces. Therefore, the matching of the linear capacitors formed by the parasitic capacitance between wires may be poorer than the matching of transistors if the spread of the threshold voltages does not dominate. Due to the smaller capacitance per square of this kind of linear capacitors, the matching is sensitive to the layout. The influence of the wire parasitic capacitance is significant.

2.3. FINITE INPUT-OUTPUT CONDUCTANCE RATIO ERRORS

The familiar square-low formula for an MOS transistor in its saturation region is given by equation (2.1). When there is a change in the drain source voltage, the drain current varies. The output conductance is defined as

$$g_{ds} = \frac{\partial I_D}{\partial V_{DS}} = \frac{\lambda I_D}{1 + \lambda V_{DS}} \approx \lambda I_D \tag{2.6}$$

In figure 2.2, we show the second-generation memory cell configuration during the two different clock phases. During one clock phase, a current flows into the memory transistor M0 that is diode-connected as shown in figure 2.2 (a); during the other clock phase, a current flows from the memory transistor M0 to a load that is the diode-connected transistor M1 as shown in figure 2.2 (b).

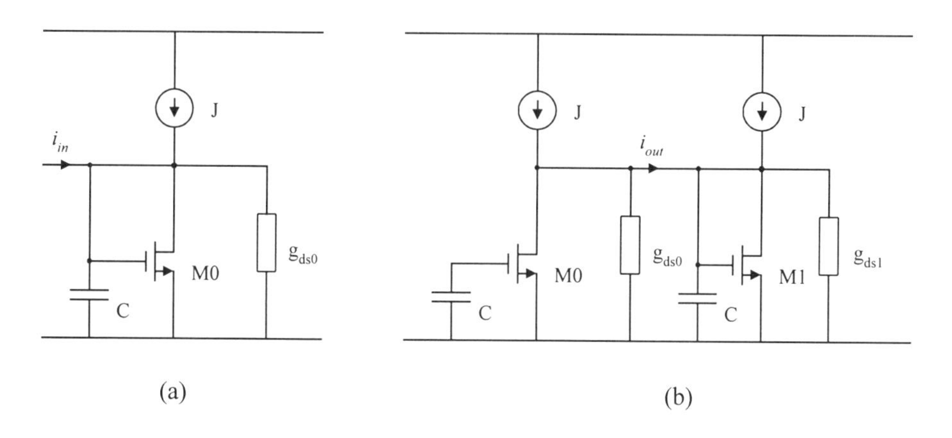

Fig. 2.2. Second-generation memory cell in its (a) input phase and (b) output phase.

Ideally, the output current should be equal to the input current delayed by half a clock period, i.e.,

$$i_{out}(n) = -i_{in}\left(n - \tfrac{1}{2}\right) \tag{2.7}$$

Due to the finite input-output conductance ratio, the output current is smaller than the ideal value. Suppose the output conductance of transistor M0 is g_{ds0} and the output conductance of transistor M1 is g_{ds1} as shown in figure 2.2. During the input phase as shown in figure 2.2 (a), the drain current of M0 is given by,

$$I_{D0}\left(n - \tfrac{1}{2}\right) = J + i_{in}\left(n - \tfrac{1}{2}\right) - g_{ds0} \cdot \Delta V_{GS0}\left(n - \tfrac{1}{2}\right) \tag{2.8}$$

where J is the quiescent bias current of the memory transistor M0, $i_{in}\left(n - \tfrac{1}{2}\right)$ is the input current, g_{ds0} is the output conductance of transistor M0 when the drain current is equal to J, and ΔV_{GS0} is the gate voltage change due to the input current. For a small input signal, ΔV_{GS0} can be approximated as

$$\Delta V_{GS0}\left(n - \tfrac{1}{2}\right) \approx \frac{i_{in}\left(n - \tfrac{1}{2}\right)}{g_{m0}} \tag{2.9}$$

where g_{m0} is the transconductance of transistor M0 when the drain current is equal to J. Therefore we have

$$I_{D0}\left(n - \tfrac{1}{2}\right) = J + \left(1 - \frac{g_{ds0}}{g_{m0}}\right) i_{in}\left(n - \tfrac{1}{2}\right) \tag{2.10}$$

During the output clock phase as shown in figure 2.2 (b), transistor M1 is diode connected, acting as a load and determining the drain potential of transistor M0. For a small input signal, the gate source voltage change of transistor M1 (i.e., the drain source voltage change of M0) is given by

$$\Delta V_{DS0}(n) = \Delta V_{GS1}(n) \approx \frac{i_{out}(n)}{g_{m1}} \approx \frac{-i_{in}\left(n - \tfrac{1}{2}\right)}{g_{m1}} \tag{2.11}$$

where g_{m1} is the transconductance of transistor M1 when the drain current is equal to the bias current J. Therefore, the output current is give by

$$
\begin{aligned}
i_{out}(n) &= J - I_{D0}(n) - g_{ds0} \cdot \Delta V_{DS0}(n) \\
&= J - I_{D0}\left(n - \tfrac{1}{2}\right) - g_{ds0} \cdot \Delta V_{DS0}(n) \\
&= -\left(1 - \frac{g_{ds0}}{g_{m0}} - \frac{g_{ds0}}{g_{m1}}\right) \cdot i_{in}\left(n - \tfrac{1}{2}\right)
\end{aligned}
\tag{2.12}
$$

Suppose that the two transistors M0 and M1 have the same size, we simplify equation (2.12) to

$$
\begin{aligned}
i_{out}(n) &= -\left(1 - 2\frac{g_{ds}}{g_m}\right) \cdot i_{in}\left(n - \tfrac{1}{2}\right) \\
&\approx -\frac{i_{in}\left(n - \tfrac{1}{2}\right)}{1 + 2\frac{g_{ds}}{g_m}} = -\frac{i_{in}\left(n - \tfrac{1}{2}\right)}{1 + \frac{2}{A_i}}
\end{aligned}
\tag{2.13}
$$

where g_m is the transconductance of transistors M0 and M1 when the drain current is equal to the quiescent bias current J, g_{ds} is the output conductance of transistors M0 and M1 when the drain current is equal to J, and A_i is defined as the input-output conductance ratio, i.e.,

$$
A_i = \frac{g_m}{g_{ds}} \tag{2.14}
$$

It is seen from equation (2.13) that the finite input-output conductance ratio reduces the output current, implying that the current transfer is not complete. If the ratio is infinite, the current transfer is complete and equation (2.13) becomes equation (2.7). The effect is exactly the same as the effect of the finite DC gain of operational amplifiers in SC circuits.

Shown in figure 2.3 is an SC unity-gain amplifier, the DC gain of the operational amplifier is A_v. During one clock phase, the input voltage is sampled by the input capacitor C_0 and the capacitor C_1 across the operational amplifier is reset as shown in figure 2.3 (a). During the other clock phase, the charge is transferred from the sampling capacitor C_0 to the capacitor C_1, generating the output voltage as shown in figure 2.3 (b).

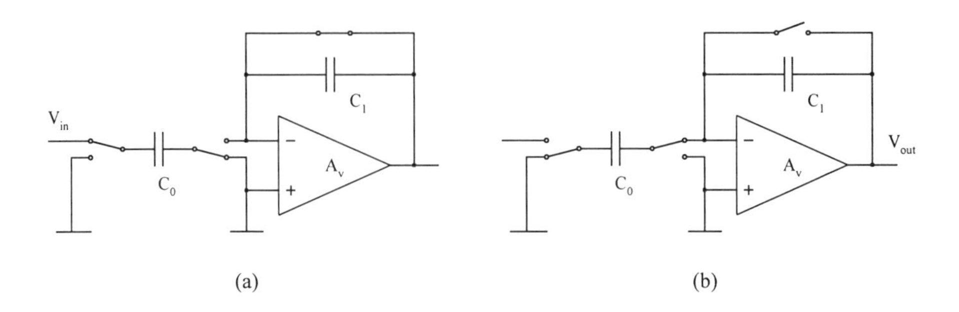

Fig. 2.3. An SC unity-gain amplifier.
(a) the sampling phase and (b) amplification phase.

Suppose the capacitor C0 and C1 have the same value, the output voltage is given [5] by

$$V_{out}(n) = \frac{V_{in}\left(n-\frac{1}{2}\right)}{1+\frac{2}{A_v}} \tag{2.15}$$

Due to the finite DC gain of the operational amplifier, the charge transfer is not complete. Comparing equations (2.13) and (2.15), we see that the effect of the finite input-output conductance ratio on SI memory cells is exactly the same as the effect of the finite DC gain of operational amplifiers on SC delayed unity-gain amplifiers.

It can be shown that the effect of the finite input-output conductance ratio on SI integrators and general SI circuits is similar to the effect of the finite DC gain of operational amplifiers on SC integrators and general SC circuits [1].

2.4. SETTLING ERRORS

The operation of SI circuits is based on charging and discharging the gate capacitance. The time of the charging and discharging process is set by the sampling clock. If at the end of the clock interval, the gate capacitance has not been charged or discharged to the final value, errors occur.

In figure 2.4 we show the first-generation SI memory cell during the input phase and its simplified small AC signal model. The switch-on resistance of the switch transistor is neglected.

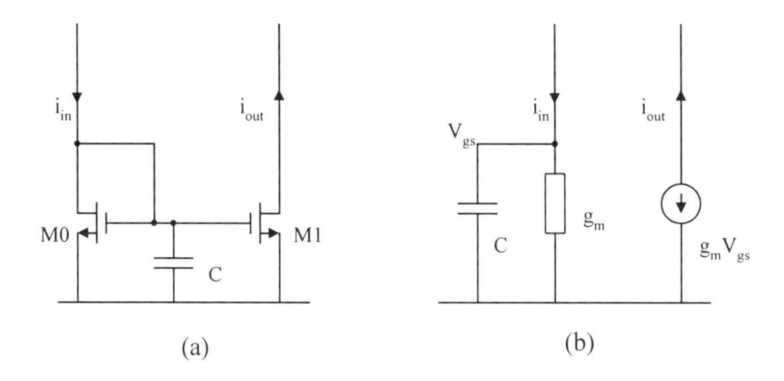

Fig. 2.4. (a) First-generation SI memory cell during the input phase and (b) its simplified small AC signal model.

Simple calculation yields the transfer function

$$H(s) = \frac{i_{out}(s)}{i_{in}(s)} = -\left(\frac{1}{1+\frac{s}{\omega_0}}\right) \tag{2.16}$$

where ω_0 is the pole radian frequency, given by

$$\omega_0 = \frac{g_m}{C} \tag{2.17a}$$

where g_m is the transconductance of transistor M0 and C is the total capacitance at the gate of transistor M0.

The pole frequency f_0 is given by

$$f_0 = \frac{\omega_0}{2\pi} = \frac{1}{2\pi} \cdot \frac{g_m}{C} \tag{2.17b}$$

The time domain response is given by

$$i_{out}(t) = -\left(1 - e^{-\omega_0 \cdot t}\right) \cdot i_{in}(t) \tag{2.18}$$

It is seen from equation (2.18) that it takes time for the SI circuit to settle to the final value depending on the pole radian frequency ω_0 formed by the gate capacitance and the transconductance. To achieve a 0.01% settling accuracy, the settling time must be 1.5 times larger than the reciprocal of the pole frequency given by equation (2.17b). Therefore, the sampling frequency must be several times smaller than the pole frequency given by equation (2.17b), otherwise large settling errors occur.

The widely used architecture of operational amplifiers in SC circuits is the single pole architecture. In figure 2.5, we show the small AC signal equivalent circuit.

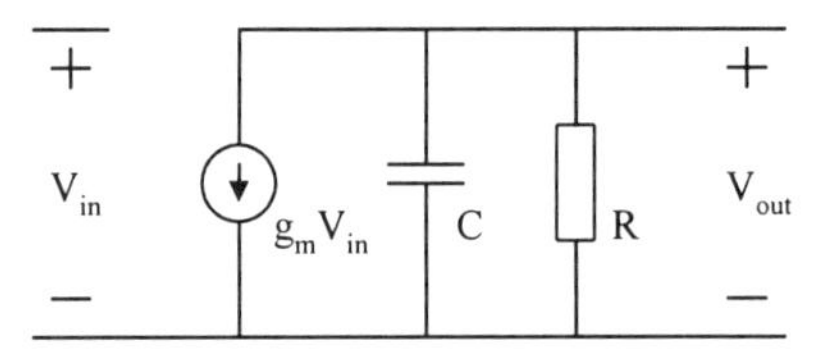

Fig. 2.5. Small AC signal equivalent circuit of a single-pole operational amplifier.

The transfer function is given by

$$H(s) = \frac{V_{out}(s)}{V_{in}(s)} = -\left(\frac{1}{\frac{1}{R} + s \cdot C}\right) \cdot g_m = -\left(\frac{1}{\frac{1}{A_0} + \frac{s}{\omega_0}}\right) \tag{2.19}$$

where R is the output resistance, C is the load capacitance, g_m is the transconductance of the input device, A_0 is the DC gain, and ω_0 is the unity-gain bandwidth.

The DC gain A_0 is given by

$$A_0 = R \cdot g_m \tag{2.20}$$

The unity-gain bandwidth is ω_0 given by

$$\omega_0 = \frac{g_m}{C} \tag{2.21}$$

If we configure the operational amplifier as a unity-gain buffer, the time-domain response is given [18] by

$$V_{out}(t) = \left(1 - \frac{A_0}{1 + A_0} e^{-\omega_0 \cdot t}\right) \cdot V_{in}(t) \approx \left(1 - e^{-\omega_0 \cdot t}\right) \cdot V_{in}(t) \tag{2.22}$$

From equations (2.18) and (2.22) it is seen that both SI and SC circuits have similar behaviors concerning the settling errors due to the limited bandwidth.

2.5. Clock Feedthrough Errors

Clock feedthrough errors are inherent in analog sampled-data MOS circuits due to the use of switches. The switches are usually MOS transistors operating alternatively in the linear and cut-off regions. Suppose that the switch is an n-type MOS transistor as shown in figure 2.6. When the voltage applied to the gate of the switch transistor goes from high to low, the switch transistor changes the operation region from the linear region to the cut-off region. The switching point is when the clock voltage is equal to the gate potential of the memory transistor M0 plus the threshold voltage of the switch transistor Ms. At this point, the impedance at the gate of the memory transistor M0 changes from low to high. Due to the parasitic gate source capacitance of the switch transistor Ms, some charge flows to the memory capacitance C, resulting an error voltage at the gate of the memory transistor. Also, when the switch transistor Ms switches from the conduction to the cut-off state, the charge established when it conducts must be completely depleted, and a part of the channel charge flows to the memory capacitor C resulting in another error voltage. The error voltages on the memory capacitance C introduce an error in the output current.

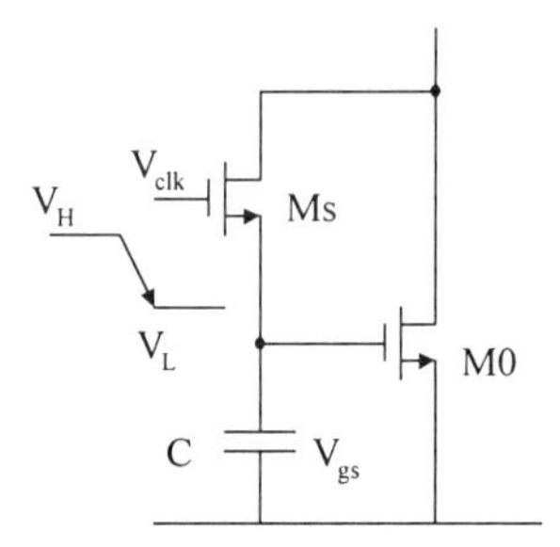

Fig. 2.6. SI memory cell with a switch transistor.

The clock feedthrough errors are dependent on many factors, e.g., the impedance at the switched nodes, operating points, switching speed, etc. The accurate analysis of the clock feedthrough errors is tedious and is of minor interest to a circuit designer. A simplified analysis usually suffices to provide some insight into the influence of clock feedthrough errors from a designer's view point.

In figure 2.7, we illustrate the capacitance inherent within an MOS transistors [19]. It consists mainly of two parts. One is the overlapping capacitance C_{ol} due to the lateral diffusion of the drain and source area, and the other is the channel capacitance C_{ch} due to the conducting layer beneath the gate when the transistor is on. Fringing field capacitance is neglected here for simplicity.

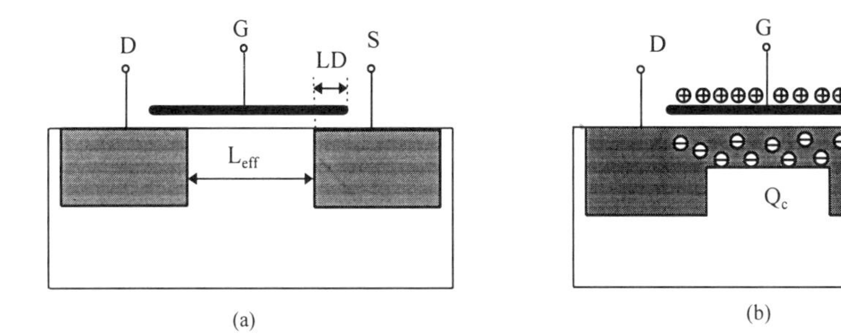

Fig. 2.7. Illustration of (a) overlapping capacitance, and (b) channel capacitance of an NMOS transistor.

The overlapping capacitance C_{ol} (at the drain or the source) is given by

$$C_{ol} = C_{ox} \cdot W_{eff} \cdot LD = C_{ox} \cdot (W - 2WD) \cdot LD \tag{2.23}$$

where C_{ox} is the gate capacitance per square, W_{eff} is the effective width, LD is the field-oxide encroachment parameter reducing the channel length from the nominal value to the effective length, and WD is the field-oxide encroachment parameter reducing the channel width from the nominal value to the effective width.

The channel capacitance C_{ch} when the transistor conducts is given by

$$C_{ch} = C_{ox} \cdot W_{eff} \cdot L_{eff} = C_{ox} \cdot (W - 2WD) \cdot (L - 2LD) \tag{2.24}$$

Now, let's see how they influence the voltage sampled at the gate of the memory transistor M0 shown in figure 2.6.

The charge injected onto the memory capacitor C due to the over-lapping capacitance is given by

$$Q_{ol} = C_{ol} \cdot (V_H - V_L) \tag{2.25}$$

where V_H is the high excursion of the clock signal applied at the gate of the switch transistor Ms and V_L is the low excursion of the clock signal applied at the gate of the switch transistor Ms.

Since the channel capacitance only exists when the transistor conducts, therefore, the total channel charge is given by

$$Q_{ch} = C_{ch} \cdot \left(V_H - V_{gs} - V_{Ts} \right) \tag{2.26}$$

where V_{gs} is the gate source voltage of the memory transistor M0, and V_{Ts} is the threshold voltage of the switch transistor Ms. (V_{gs} + V_{Ts}) determines the switching point as shown in figure 2.6.

The threshold voltage V_{Ts} of the switch transistor Ms is a function of its source bulk voltage (which is equal to the gate source voltage V_{gs} of the memory transistor M0) and is given [4] by

$$V_{Ts} = V_{T0} + \gamma \cdot \left\{ \sqrt{2|\Phi_F| + V_{gs}} - \sqrt{2|\Phi_F|} \right\} \approx V_{T0} + \frac{\gamma}{3} V_{gs} \tag{2.27}$$

where V_{T0} is the threshold voltage of the switch transistor Ms with a zero source bulk voltage, γ is bulk-threshold parameter, Φ_F is the Fermi potential, and V_{gs} is the gate source voltage of the memory transistor M0.

When the switch transistor Ms cuts off, all the charge due to the overlapping capacitance and a portion of the charge due to the channel capacitance flow onto the memory capacitor C. The resulting error voltage at the gate of the memory transistor is given by

$$\Delta V_{gs} = \frac{Q_{ol} + \alpha \cdot Q_{ch}}{C} \tag{2.28}$$

where α determines the portion of the channel charge that flows to the memory capacitor C, dependent on many factors, such as the nodal impedance, the switching speed, etc. The reasonable value is usually between 0.5 and 1.

Therefore, the error current Δi can be approximated by

$$\Delta i = g_m \cdot \Delta V_{gs} \tag{2.29}$$

where g_m is the transconductance of the memory transistor with the drain current equal to $J + i$ (where J is the quiescent bias current and i is the input current), and is given by

$$\begin{aligned} g_m &\approx \sqrt{\mu C_{ox}(J+i)\frac{W_{eff}}{L_{eff}}} \approx \sqrt{\mu C_{ox} J \frac{W_{eff}}{L_{eff}}} \cdot \left(1 + \frac{1}{2} \cdot \frac{i}{J}\right) \\ &= g_{m0} \cdot \left(1 + \frac{1}{2} \cdot \frac{i}{J}\right) \end{aligned} \tag{2.30}$$

where μ is the charge mobility, C_{ox} is the gate capacitance per square, J is the quiescent bias current, i is the input current, W_{eff} is the effective transistor width, L_{eff} is the effective length, and g_{m0} is the transconductance of the memory transistor M0 with the drain current equal to the quiescent bias current J (i.e., zero input current). In the above approximate, we assumed that the input current i is much smaller than the quiescent bias current J and therefore the Taylor series expansion holds.

The gate source voltage of the memory transistor M0 is given by

$$\begin{aligned} V_{gs} &= \sqrt{\frac{(J+i)}{\frac{1}{2}\mu C_{ox}\frac{W_{eff}}{L_{eff}}}} + V_{T0} \\ &= \sqrt{\frac{J}{\frac{1}{2}\mu C_{ox}\frac{W_{eff}}{L_{eff}}}} \cdot \left(1 + \frac{1}{2} \cdot \frac{i}{J}\right) + V_{T0} \\ &= \left(V_{gs0} - V_{T0}\right) \cdot \left(1 + \frac{1}{2} \cdot \frac{i}{J}\right) + V_{T0} \end{aligned} \tag{2.31}$$

where μ is the charge mobility, C_{ox} is the gate capacitance per square, J is the quiescent bias current, i is the input current, W_{eff} is the effective transistor width, L_{eff} is the effective length, V_{T0} is the threshold voltage of transistor M0 with a zero source bulk voltage, and V_{gs0} is the gate source voltage of the memory transistor M0 with the drain current equal to the quiescent bias current J (i.e., zero input current). In the above approximate, we assumed that the input current i is much smaller than the quiescent bias current J and therefore the Taylor series expansion holds. We also made the simplification that the threshold voltages of the memory transistor M0 and the switch transistor Ms are the same when their source bulk voltages are zero.

Combining equations (2.23) ~ (2.31) and neglecting higher-order components (because the input current i is much less than the quiescent bias current J), we have

$$\Delta i \approx \omega_0 \cdot \left\{ k_i + k_d \cdot \frac{i}{J} \right\} \tag{2.32}$$

where ω_0 is the pole radian frequency of the memory cell at the quiescent condition, given by

$$\omega_0 = \frac{g_{m0}}{C} \tag{2.33}$$

and k_i is the coefficient of the signal independent part and k_d is the coefficient of the signal dependent part. They are given, respectively, by

$$k_i = C_{ol} \cdot (V_H - V_L) + \alpha \cdot C_{ch} \cdot \left\{ V_H - \frac{3+\gamma}{3} \cdot V_{gs0} - V_{T0} \right\} \tag{2.34}$$

and

$$\begin{aligned} k_d &= \frac{1}{2} k_i - \alpha \cdot C_{ch} \cdot \frac{3+\gamma}{6} \cdot (V_{gs0} - V_{T0}) \\ &= \frac{1}{2} C_{ol} \cdot (V_H - V_L) + \frac{1}{2} \alpha \cdot C_{ch} \cdot \left\{ V_H - 2 \cdot \frac{3+\gamma}{3} \cdot V_{gs0} + \frac{\gamma}{3} \cdot V_{T0} \right\} \end{aligned} \tag{2.35}$$

The error current due to the clock feedthrough is directly proportional to the cut-off frequency of the memory cell. Therefore, higher speed SI circuits suffer more from the clock feedthrough errors.

There are two major contributors to the error current, one is signal independent and the other is signal dependent. Both of them are dependent on the overlapping and channel capacitance of the switch transistor, the clock voltages (V_H and V_L), the clock transition(via the parameter α), and the DC operating points of the memory cell.

Unlike other derivations (e.g. [1]) where the clock feedthrough error of fully-differential SI circuits is independent of the clocking and the overlapping capacitance of the switch transistor, from equations (2.33)~(2.35) it is seen that even with fully differential architecture (signal-independent part disappears), the clock feedthrough error is still heavily dependent on the clocking and the overlapping capacitance as well as on the DC operating points of the memory cell. The reason lies in the fact that the transconductance of the memory cell is dependent on the input current. In a fully differential SI circuit, the currents flowing into the two branches have a different direction making the transconductance of the two branches differ significantly. The assumption of the same transconductance is somehow over simplified. In the above derivation, the dependence of the transconductance on the input current is considered.

The clock feedthrough error is heavily input signal dependent as can be seen from equations (2.33)~(2.35). When the input current increases in respect to the quiescent bias current, the error current is more dependent on the input signal. Therefore, SI circuits usually have large distortions.

When we sample a voltage onto a capacitor C as shown in figure 2.8, we also have clock feedthrough errors.

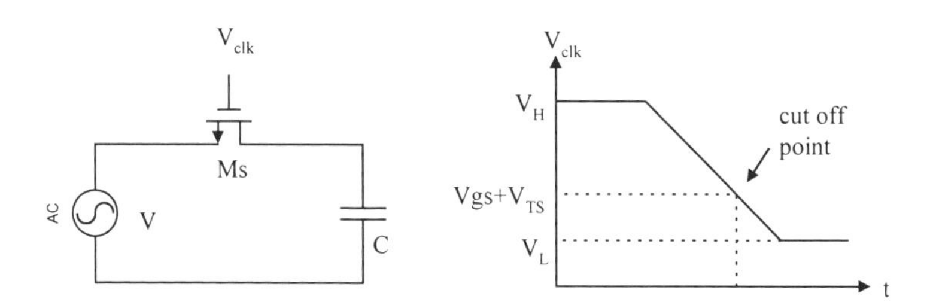

Fig. 2.8. Clock feedthrough error in SC circuits.

The error voltage is given by

$$\Delta V = \frac{Q_{ol} + \alpha \cdot Q_{ch}}{C} \tag{2.36}$$

where Q_{ol} is the total charge due to the overlapping capacitance of the switch transistor Ms given by equation (2.25), Q_{ch} is the total channel charge due to the channel capacitance of the switch transistor Ms given by equation (2.26), and α determines the portion of the channel charge that flows to the sampling capacitor C, dependent on the switching speed.

By using equations (2.25)~(2.27), we have

$$\Delta V = \frac{1}{C}\left\{C_{ol}\left(V_H - V_L\right) + \alpha C_{ch}\left(V_H - V_{T0}\right)\right\} - \frac{1}{C}\alpha C_{ch} \cdot \frac{3+\gamma}{3} V \tag{2.37}$$

where V is the input voltage being sampled by the capacitor C.

Notice that the clock feedthrough error in SC circuits also has both signal independent and signal dependent part. The signal dependent part (the second term) is independent of the clock voltages and the overlapping capacitance of the switch transistor. Therefore the clock feedthrough error in fully differential SC circuits is independent of clock voltages and overlapping capacitance of the switch transistor Ms. SC circuits usually have lower signal dependent clock feedthrough errors due to the absence of the signal dependent transconductance.

For SC circuits, the signal dependent clock feedthrough error can be eliminated to the first order by clocking. An example is shown in figure 2.9.

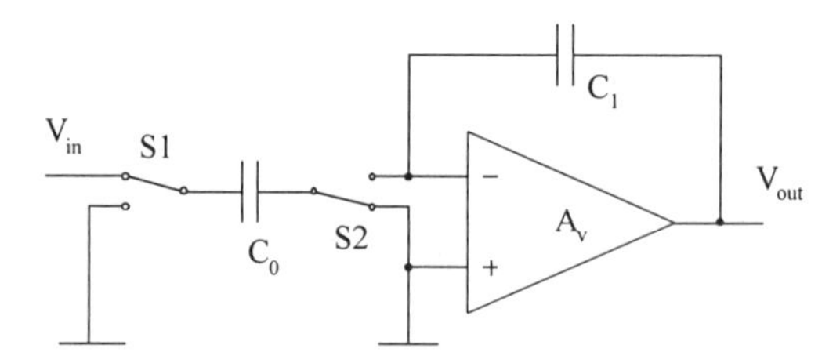

Fig. 2.9. Illustration of signal dependent clock feedthrough free SC circuits.

Suppose the SC circuit shown in figure 2.9 is controlled by a non-overlapping clock. On one clock phase, the input voltage V_{in} is sampled onto the sampling capacitor C_0 by two switches S1 and S2. If S1 and S2 are controlled by the same clock signal, then a signal dependent clock feedthrough error results as shown in equation (2.37). However, if we open switch S2 slightly earlier than S1, we can eliminate the signal dependent clock feedthrough.

When the switch S2 opens while the switch S1 is still closed, a clock feedthrough error is introduced onto the sampling capacitor C_0. Since the voltage is always zero, it does not introduce a signal dependent clock feedthrough error. After switch S2 is opened switch S1 opens. Since capacitor C_0 is already floating, the charge due to the clock feedthrough cannot be injected onto capacitor C_0 to the first order. Therefore, no signal dependent clock feedthrough error is injected onto the sampling capacitor C_0.

In SI circuits, a similar operation arrangement is not possible due to the inherent nature. Therefore, SI circuits have larger distortions than SC circuits due to the signal dependent clock feedthrough errors.

2.6. Drain-Gate Capacitive Coupling Errors

Another error source in SI circuits is due to the drain-gate capacitive coupling. Referring to figure 2.2, the drain voltage change has a significant influence on the accuracy of SI circuits due to the finite input-output conductance ratio. Also the drain voltage change can be directly coupled onto the memory capacitor C due to the parasitic drain-gate capacitance C_{dg}.

If we do not consider the effect of switching, the existence of the drain-gate capacitance C_{dg} further increases the output conductance of the memory transistor [1]. The total output conductance is given [1] by

$$g_0 = g_{ds} + \frac{C_{dg}}{C + C_{dg}} g_m \tag{2.38}$$

where g_{ds} is the output conductance of the memory transistor M0, C is the memory capacitor, g_m is the transconductance of the memory transistor M0, and C_{dg} is the drain-gate parasitic capacitance. The input-output conductance ratio according to equation (2.14) is given by

$$A_i = \frac{g_m}{g_0} = \frac{1}{\frac{g_{ds}}{g_m} + \frac{C_{dg}}{C + C_{dg}}} \tag{2.39}$$

The finite value of the input-output conductance ratio introduces errors in SI circuits as discussed in Chapter 2.2.

If the output-input conductance ratio of a memory cell (g_{ds}/g_m) is made small (e.g., by increasing the transconductance g_m to increase the speed), then the dominate factor is the capacitive coupling due to the drain-gain capacitance and equation (2.39) becomes

$$\begin{aligned} A_i &= \frac{g_m}{g_0} = \frac{1}{\frac{g_{ds}}{g_m} + \frac{C_{dg}}{C + C_{dg}}} \approx \frac{C + C_{dg}}{C_{dg}} \\ &\approx \frac{C_{ox} \cdot W \cdot L}{C_{ox} \cdot W \cdot LD} = \frac{L}{LD} \end{aligned} \tag{2.40}$$

where C_{ox} is the gate capacitance per square, W is the width of the memory transistor, L is the length of the memory transistor, LD is the field-oxide encroachment parameter reducing the channel length from the nominal value to the effective length as shown in figure 2.7 (a). The value can be as low as 10 for short channel devices in order to have a high speed operation. The value can be increased by increasing the length at the cost of speed. Another way to increase the value is to add an extra capacitor at the gate, also at the cost of speed.

To reduce the error due to the drain-gate capacitive coupling, circuit techniques such as the cascoding and grounded-gate amplifier techniques can be used as these techniques are used to reduce the finite input-output conductance ratio error. However, these techniques are not so effective to reduce the error due to the drain-gate capacitive coupling when large transient glitches are present.

When we switch the second-generation SI circuits, e.g., the memory cell shown in figure 2.2, the drain potential of the memory cell may have a large spike before it changes from the input drain potential to the output drain potential. This spike couples onto the memory capacitance via the drain-gate capacitance. Since the drain-gate capacitance is dependent on the drain potential, a non-linear error is introduced even if the drain potential eventually settles down to the right value [20]. To reduce this effect, a special clocking is needed [17, 18, and 21].

2.7. NOISE

There are different types of noise in integrated analog circuits, and in MOS analog circuits the dominant noise sources are flicker nose and thermal noise [22].

The noise power spectral density of flicker noise referred to the gate of an MOS transistor is given [22] by

$$\frac{\overline{v_{nf}^2}}{\Delta f} = \frac{K_f}{WLC_{ox}f} \tag{2.41}$$

where K_f is the flicker noise coefficient depending on the process, W is the width of the MOS transistor, L is the length of the MOS transistor, and C_{ox} is the gate capacitance per square.

It is seen from equation (2.41) that the power spectral density of flicker noise is inversely proportional to the frequency. This is why flicker noise is also referred to as $1/f$ noise.

The noise power spectral density of thermal noise referred to the gate of an MOS transistor is given [22] by

$$\frac{\overline{v_{nt}^2}}{\Delta f} = 4kT \cdot \frac{2}{3} \cdot \frac{1}{g_m} \tag{2.42}$$

where k is the Boltzmann constant, T is the absolute temperature, and g_m is the transconductance of the MOS transistor.

From equation (2.42) it is seen that the thermal noise spectral density is independent of the frequency, and it is also referred to as white noise.

Noise can also be referred to the drain current of an MOS transistor. The current noise power spectral density is determined by the voltage noise power spectral density and the transconductance. The current noise power spectral density of flicker noise is given by

$$\frac{\overline{i_{nf}^2}}{\Delta f} = \frac{\overline{v_{nf}^2}}{\Delta f} \cdot g_m^2 = \frac{g_m^2 \cdot K_f}{WLC_{ox} f} \tag{2.43}$$

The current noise power spectral density of thermal noise is given by

$$\frac{\overline{i_{nt}^2}}{\Delta f} = \frac{\overline{v_{nt}^2}}{\Delta f} \cdot g_m^2 = 4kT \cdot \frac{2}{3} \cdot g_m \tag{2.44}$$

Flicker noise is usually considerably larger than thermal noise when the bandwidth is in the kHz range. Therefore, for audio applications where low noise is demanded, flicker noise needs to be reduced by using circuit techniques such as the chopper stabilization and the correlated double sampling techniques[4]. When the signal bandwidth is in MHz range, it is usually thermal noise that limits the dynamic range.

The correlated double sampling is actually inherent in the second-generation SI circuits. Shown in figure 2.10 is the second-generation SI memory cell. The noisy memory cell is modeled by a noiseless memory transistor M0 and a noise source i_n. Notice that the noise source i_n can even represent the noise contribution from the current source.

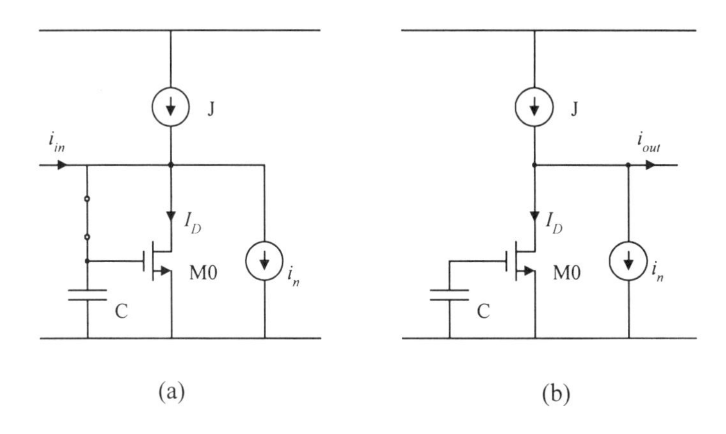

Fig. 2.10. Illustration of the correlated double sampling in the second-generation SI memory cell. (a) during the input clock phase, and (b) during the output clock phase.

At the end of the input clock phase, the drain current in the noiseless memory cell is given by

$$I_D(n-\tfrac{1}{2}) = J + i_{in}\left(n-\tfrac{1}{2}\right) - i_n\left(n-\tfrac{1}{2}\right) \tag{2.45}$$

where J is the quiescent bias current. i_{in} is the input current, and i_n is the noise current.

At the end of the output clock phase, the drain current in the noiseless memory cell is given by

$$I_D(n) = J - i_{out}(n) - i_n(n) \tag{2.46}$$

where i_{out} is the output current.

Due to the fact that the gate voltage is held by the gate capacitance C, we have

$$I_D(n-\tfrac{1}{2}) = I_D(n) \tag{2.47}$$

Therefore, the output current is given by

$$i_{out}(n) = -i_{in}\left(n - \tfrac{1}{2}\right) - \left\{ i_n\left(n - \tfrac{1}{2}\right) - i_n(n) \right\} \tag{2.48}$$

Since flicker noise only has low frequency components, if the sampling frequency is much higher than the corner frequency of the flicker noise, we have

$$i_n\left(n - \tfrac{1}{2}\right) \approx i_n(n) \tag{2.49}$$

Therefore equation (2.48) becomes

$$i_{out}(n) = -i_{in}\left(n - \tfrac{1}{2}\right) \tag{2.50}$$

From equation (2.50) it is seen that the influence of the low frequency noise is eliminated due to the correlated double sampling.

In Chapter 1.2, we compared the noise in SI and SC circuits. Here we recapitulate the comparison.

Neglecting the contribution of the low frequency noise such as flicker noise, we only consider thermal noise. The current noise power spectral density of the memory cell shown in figure 2.10 is given by

$$\frac{\overline{i_n^2}}{\Delta f} = \frac{8}{3} \cdot k \cdot T \cdot \left(g_{m0} + g_{mJ}\right) \tag{2.51}$$

where k is the Boltzmann constant, T is the absolute temperature, g_{m0} is the transconductance of the memory cell M0, and g_{mJ} is the transconductance of the transistor forming the current source J.

The noise bandwidth of a single pole system such as the memory cells shown in figure 2.10 is given by

$$BW_n = \frac{\pi}{2} \cdot f_{pole} = \frac{\pi}{2} \cdot \frac{1}{2\pi\tau} = \frac{g_{m0}}{4C} \tag{2.52}$$

where f_{pole} is the pole frequency, τ is the time constant, g_{m0} is the transconductance of the memory transistor M0, and C is the total capacitance at the gate of the memory transistor M0.

Therefore, the total current noise power is the product of the noise power spectral density and the noise bandwidth, i.e.,

$$\overline{i_{rms}^2} = \frac{\overline{i_n^2}}{\Delta f} \cdot BW_n = \frac{kT}{C} \cdot \left\{ \frac{2}{3} g_{m0}^2 \cdot \left[1 + \frac{g_{mJ}}{g_{m0}} \right] \right\} \tag{2.53}$$

If we refer the current noise to the gate of the memory transistor M0, we have the total voltage noise power given by

$$\overline{v_{rms}^2} = \frac{\overline{i_{rms}^2}}{g_{m0}^2} = \frac{kT}{C} \cdot \left\{ \frac{2}{3} \cdot \left[1 + \frac{g_{mJ}}{g_{m0}} \right] \right\} \approx \frac{kT}{C} \tag{2.54}$$

The approximation takes into consideration that the transconductance of the current source transistor is usually smaller or in the same range as the transconductance of the memory transistor M0.

When we operate the SI memory cells with a sampling clock, the sampling operation does not change the total noise power but only redistributes it. If we neglect the noise contribution from the switch-on resistance of the switch transistors [1], the total noise power is in the same range as given by equation (2.54).

When we sample a signal, all the thermal noise folds into the signal band for SC circuits as well. This imposes a fundamental limitation in SC circuits. When sampling a voltage signal on a capacitor C, the total noise power is given by,

$$\overline{v_{rms}^2} = \frac{kT}{C} \tag{2.55}$$

where k is the Boltzmann constant, T is the absolute temperature, and C is the sampling capacitance. Notice the total noise power is independent of the switch-on resistance and the sampling frequency.

We see that the fundamental limitation of thermal noise is the same for both SI (equation (2.54)) and SC circuits (equation (2.55)). Usually, the parasitic capacitor in SI circuits is considerably smaller than the sampling capacitor in SC circuits, and therefore, SI circuits are more noisy than SC circuits.

2.8. Summary

In this chapter, we have detailed nonidealities in SI circuits and compared them to the SC counterparts, including the mismatch errors, the finite input-output conductance ratio errors, the settling errors, the clock feedthrough errors, the drain-gate parasitic capacitve coupling errors, and the noise errors.

The matching property of SI circuits is essentially the same as the matching property of current mirrors, depending on both the process and the parasitic resistance in the actual layout. Due to the mismatch in the threshold voltage, the matching of transistors in SI circuits is usually poorer than the matching of linear capacitors in SC circuits.

The finite input-output conductance ratio of memory cells inflicts an incomplete current transfer in SI circuits just as the finite DC gain of operational amplifiers in SC circuits inflicts an incomplete charge transfer.

The settling error due to the pole frequency in SI circuits is similar to the settling error due to the unity-gain bandwidth of operational amplifiers in SC circuits, though the pole frequency of SI circuits is usually higher than the unity-gain bandwidth of operational amplifiers in SC circuits. To reduce the settling error, the clock frequency must be reduced in respect to the pole frequency.

The clock feedthrough error in SI circuits is dependent on many factors such as the clock voltages, the switching speed, the overlapping and channel capacitance, the operating points, and the input current. It contains both signal independent and signal dependent parts. The clock feedthrough error is usually larger in SI circuits than in SC circuits due to the large signal dependent portion. One of major reasons is the strong dependency of the transconductance on the input current (and this is usually neglected in the SI literature).

The drain-gate parasitic capacitance does not only increases the output conductance but also introduces distortion in the second-generation SI circuits due to the large transient current spikes that may be coupled onto the memory capacitor via the nonlinear drain-gate parasitic capacitance.

Inherent in the second-generation SI circuits, flicker noise is suppressed due to the correlated double sampling. Thermal noise imposes fundamental limitation in both SI circuits and SC circuits, thought the noise in SI circuits is usually larger than that in SC circuits due to the smaller value of the sampling capacitance.

In the next chapter, we will present practical SI circuits and circuit techniques that can reduce the nonidealities we have discussed in this chapter.

Chapter III. Practical SI Circuits

3.1. Introduction

SI circuits, as current-mode circuits feature inherently wide bandwidth and suitability for low-voltage operations and are completely compatible with the digital CMOS process [1, 17]. However, as described in Chapter II, SI circuits based on basic memory cells deviate from the ideal behavior, and all errors increase with the bandwidth. Therefore circuit techniques are needed to make SI circuits more competitive.

In this chapter, we will present SI circuits and circuit techniques which enable us to utilize the SI technique in a system such as an oversampling A/D converter. Measurement results of the practical SI circuits will be included as well.

3.2. First-Generation SI Circuits with Clock Feedthrough Compensated

The transmission error due to the finite input-output conductance ratio can be reduced to an acceptable level, if we increase the input conductance or decrease the output conductance [1]. One of the most troublesome errors is the error introduced by the clock charge injection, also known as the clock feedthrough error, and the error is signal dependent as given by equations (2.32)~(2.35).

Dummy switches are extensively used for SC circuits [23]. For SI circuits, the compensation using dummy switches is sensitive to the impedance difference at the drain and source of the switch transistor and to the clocking edge timing [19, 23, 24]. These effects are characterized in the coefficient α in Chapter 2.5.

A compensation technique was proposed to reduce the clock feedthrough error for the first-generation SI circuits [25] and was successfully applied to a filter design [26] and an oversampling delta-sigma modulator design [27].

The improved clock feedthrough compensated first-generation SI memory cell [26, 27] is shown in figure 3.1.

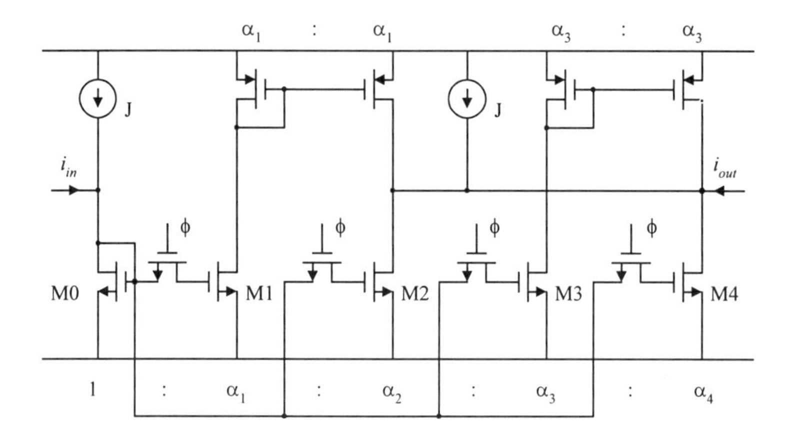

Fig. 3.1. First-generation SI memory cell with clock feedthrough compensated.

Assume that all switch transistors are identical and that transistor dimension ratios are as indicated in the figure. By some mathematical manipulations [19], we have

$$i_{out} = \left[(\alpha_4 - \alpha_3) + (\alpha_2 - \alpha_1)\right] \cdot i_{in} + \left[\left(\frac{1}{\alpha_4} - \frac{1}{\alpha_3}\right) + \left(\frac{1}{\alpha_2} - \frac{1}{\alpha_1}\right)\right] \cdot \frac{\beta_0}{2} \cdot V_c^2 \tag{3.1}$$

where β_0 is the gain factor, or the transconductance factor of M0 given by equation (2.2), and V_C is the clock feedthrough error voltage at the gate of M0.

Unity gain memory cells require

$$(\alpha_4 - \alpha_3) + (\alpha_2 - \alpha_1) = 1 \tag{3.2}$$

The condition for the complete cancellation of the clock feedthrough error is given by

$$\left(\frac{1}{\alpha_4} - \frac{1}{\alpha_3}\right) + \left(\frac{1}{\alpha_2} - \frac{1}{\alpha_1}\right) = 0 \tag{3.3}$$

If equations (3.2) and (3.3) are satisfied, the SI memory cell shown in figure 3.1 will theoretically cancel both the signal independent and signal dependent clock feedthrough errors, thus realizing a clock-feedthrough-free current memory cell. However, due to the process variation, equations (3.2) and (3.3) can not be accurately satisfied, and the clock feedthrough errors will not be completely canceled. Nevertheless, simulation results indicate a substantial reduction in the clock feedthrough errors compared with other approaches [19, 25].

In figure 3.2, we show the measured power spectrum of the memory cell of figure 3.1. Cascode transistors are used for both memory transistors and current sources to decrease the output conductance, and the supply voltage is 3.3 V. When the input is a 14-μA 10-kHz sinusoid and the clock frequency is 1 MHz, the measured total harmonic distortion (THD) is about - 65 dBc. (The bias current is about 150 μA.)

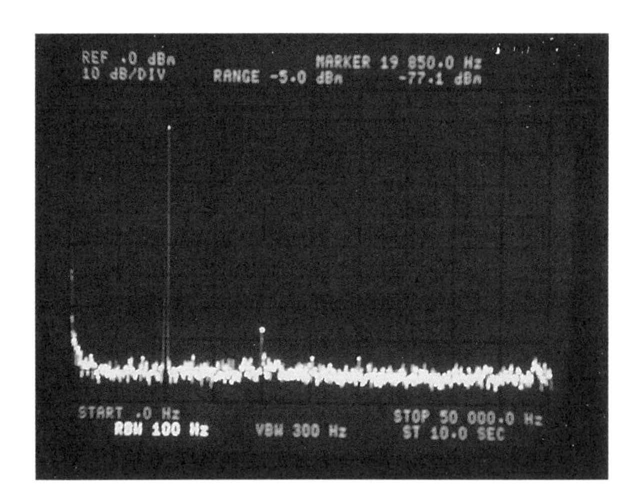

Fig. 3.2. Measured power spectrum of the SI memory cell of Fig. 3.1. THD is about - 65 dBc.

3.3. Low-Voltage Fully Differential SI Circuits

To reduce errors in SI circuits, fully differential structures are the good candidate. Being able to obtain all the advantages of fully differential structures, fully differential SI circuits have an added advantages, i.e., reduced clock feedthrough error [1, 28]. As discussed in Chapter 2.5, all the signal independent clock feedthrough error is canceled by using a fully differential architecture. Therefore, clock feedthrough errors can be reduced substantially by using fully differential architectures.

As SI memory cells have a single input and often multiple outputs, it can be advantageous to increase the input conductance in the sense that the enhancement circuit can be shared by many blocks. Obviously, the transconductance of an MOS transistor is too low to be used to create a virtual ground at the input. Some kind of feedback using a high gain amplifier is needed. Then the effective input impedance can be reduced by the gain of the amplifier employed. We can either use an operational amplifier or simply a grounded gate amplifier (GGA) for this purpose [1].

The circuit technique employing an operational amplifier [1] may result in a poor settling behavior due to the parasitic poles of the switch and the operational amplifier. An alternative arrangement having a better settling behavior is to employ a grounded-gate amplifier to create a 'virtual ground' at the drain of the memory transistor [1, 28].

In figure 3.3 (a), we show a conventional fully differential SI memory cell [28], while in figure 3.3 (b), we show the low-voltage fully differential SI memory cell [29, 30].

The GGA consists of the grounded-gate transistor T_G and its biasing transistors. By using the GGA, the input conductance of the memory transistor T_1 (and T_2) is increased to

$$g_{in} = g_m \cdot A_{GGA} \approx g_m \cdot \frac{g_{mG}}{g_{dsG}} \tag{3.4}$$

where g_m is the transconductance of the memory transistor T_1 (and T_2), A_{GGA} is the gain of the GGA, g_{mG} is the transconductance of the grounded-gate transistor T_G, and g_{dsG} is the output conductance of the grounded-gate transistor T_G. (The output conductance of the cascode current source for T_G is relatively small and is neglected.)

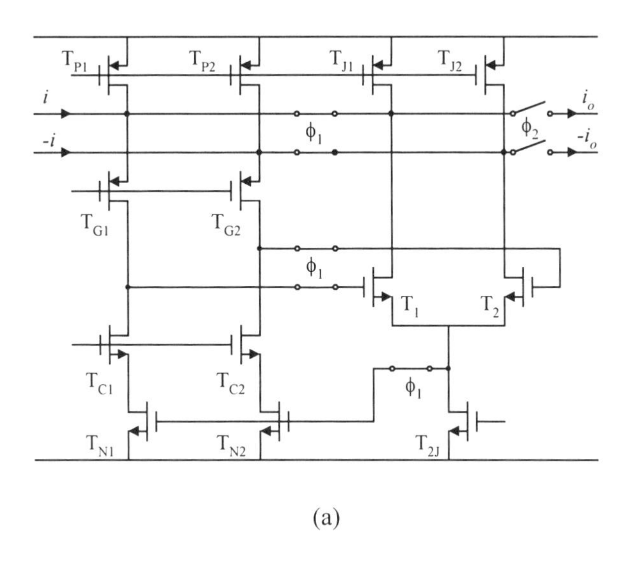

(a)

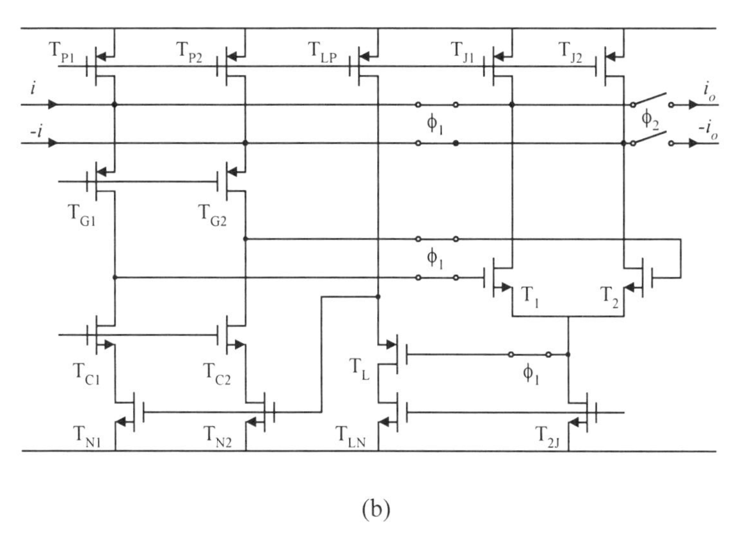

(b)

Fig. 3.3. (a) Conventional fully differential SI memory cell and (b) low-voltage fully differential SI memory cell.

The input conductance can be increased by 20 ~ 40 dB for a practical design. The output conductance of the memory cells shown in figure 3.3 is the summation of the output conductance of the memory transistor T_1 (and T_2) and the output conductance of the bias transistor T_J. Therefore, the error due to the finite input-output conductance ratio can be reduced substantially (by the gain of the GGA).

The main difference between the memory cells of figures 3.3 (a) and 3.3 (b) is the use of the level shifter in figure 3.3 (b) consisting of a level shifting transistor T_L and its biasing transistors T_{LN} and T_{LP}. The gate voltages of the common-mode feedback transistors T_{N1} and T_{N2} are thus shifted by the value of V_{gsL} (where V_{gsL} is the gate source voltage of T_L at the quiescent condition). Except for this difference, these two memory cells would be identical. Due to this very difference, significant improvements result.

To ensure the proper operation, every transistor should operate in its saturation region. Referring to figure 3.3 (a), the drain voltage of the biasing transistor T_{2J} is not set by its saturation voltage but the voltage necessary to turn on T_{N1} and T_{N2}. This larger than necessary drain voltage of T_{2J} eliminates the possibility of a low-voltage operation. An inspect into the circuit of figure 3.3 (a) reveals that the critical path setting the lower limit of the power supply voltage is through T_P, T_G, T, and T_N. The power supply voltage for the memory cell of figure 3.3 (a) should be

$$\begin{aligned} V_{dd} \geq & \left|\left(V_{gs}-V_T\right)_P\right| + \left|\left(V_{gs}-V_T\right)_G\right| + V_T + \left(V_{gs}-V_T\right)\sqrt{1+m_i} \\ & + V_{TN} + \left(V_{gs}-V_T\right)_N \sqrt{1+m_c} \end{aligned} \tag{3.5}$$

where $\left(V_{gs}-V_T\right)_P, \left(V_{gs}-V_T\right)_G, \left(V_{gs}-V_T\right), \text{and} \left(V_{gs}-V_T\right)_N$ are the quiescent saturation voltages of T_P, T_G, T, and T_N respectively, V_T and V_{TN} are the threshold voltages of T and T_N respectively, m_i is the differential signal modulation index (i/J) (J is the quiescent drain current of T), and m_c is the common-mode signal modulation index (i_{cm}/I_N) (I_N is the quiescent drain current of T_N).

Referring to figure 3.3 (b), due to the use of the level shifter, the drain voltage of T_{2J} only needs to keep T_{2J} in the saturation region. The critical path limiting the power supply voltage is through T_P, T_G, T, and T_{2J}. The power supply voltage for the memory cell of figure 3.3 (b) should be

$$V_{dd} \geq \left|\left(V_{gs}-V_T\right)_P\right| + \left|\left(V_{gs}-V_T\right)_G\right| + V_T + \left(V_{gs}-V_T\right)\sqrt{1+m_i} + \left(V_{gs}-V_T\right)_{2J} \tag{3.6}$$

where $\left(V_{gs}-V_T\right)_P, \left(V_{gs}-V_T\right)_G, \left(V_{gs}-V_T\right), \text{and} \left(V_{gs}-V_T\right)_{2J}$ are the quiescent saturation voltages of T_P, T_G, T, and T_{2J} respectively, V_T is the threshold voltages of T , m_i is the differential signal modulation index (i/J).

Comparing equations (3.5) and (3.6) it is seen that the lower limit of the power supply voltage is reduced approximately by the amount of one threshold voltage.

By cascading two memory cells and sharing the GGAs and the level shifter, we have a delay line. In figure 3.4 we show the measured power spectrum of the delay line with a single 3.3-V power supply.

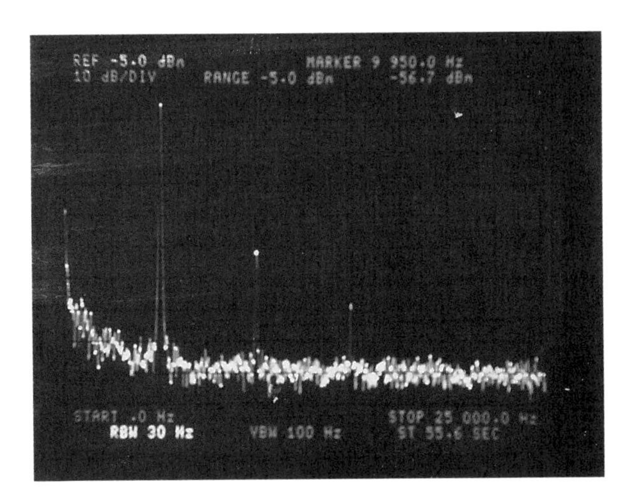

Fig. 3.4. Measured power spectrum of the delay line consisting of two SI memory cells of Fig. 3.3. THD < - 45 dBc.

When the input is a 8-μA 5-kHz sinusoid and the clock frequency is 5 MHz, the measured THD is less than - 45 dBc. (The bias current is about 20 μA.)

It is seen in figure 3.4 that the second harmonic dominates even though the architecture is fully differential. The reason lies in the fact that the transconductance of the differential pair T_1 and T_2 differ when there is an input current. The current in one branch (e.g., memory transistor T_1) increases and the current in the other branch (memory transistor T_2) decreases, which makes the transconductance differ. This makes the two branches have a different settling time, different input-output conductance ratio, and different clock feedthrough. In one word, the difference in the transconductance introduces large errors, making the fully-differential architecture less effective for SI circuits than for SC circuits. Nevertheless, fully-differential SI circuits tend to cancel all the signal independent errors, offering advantages over the single-ended counterparts at the cost of one extra branch and common-mode control circuits.

The memory circuit shown in figure 3.1 has a lower distortion than the circuit shown in figure 3.3 (b) even though the memory cell of figure 3.1 is a single-ended architecture. The major reasons are 1) the first-generation SI circuits usually have lower distortions than the second-generation SI circuits due to smaller transient switching glitches [31]; 2) the clock feedthrough error is compensated; 3) smaller input index; and 4) smaller bandwidth.

3.4. From CMFB to CMFF

To utilize the advantages of fully differential SI circuits, the common-mode feedback (CMFB) as shown in figure 3.3 is needed to control the common-mode components.

The traditional CMFB has the following drawbacks while processing the common-mode components: 1) nonlinearity due to the use of inherent voltage-to-current and current-to-voltage conversions; and 2) speed limitation due to the use of the feedback loop. Also noted is the limitation of the reduction in the power supply voltage due to the larger than necessary drain voltage for the common-mode sense transistors [1, 28], though a level shifting can circumvent it [29, 30] as discussed in Chapter 3.3.

To eliminate all the drawbacks, the general common-mode feedforward (CMFF) [16, 17] technique as illustrated in figure 3.5 can be used. It processes the common-mode components in a truly current-mode fashion.

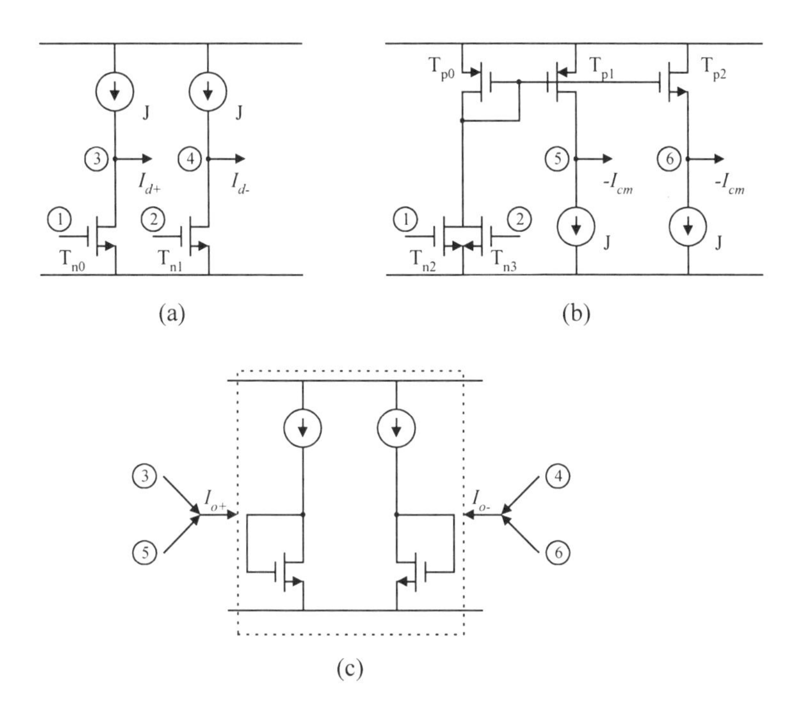

Fig. 3.5. Illustration of the common-mode feedforward.

(a) a model of a general current-mode circuit block during its output phase. J is the biasing current, I_{d+} and I_{d-} are the differential output currents. Transistors T_{n0} and T_{n1} are matched.

(b) a way of generating the common-mode current. Transistors T_{n2} and T_{n3} have the half size of transistors T_{n0} and T_{n1}. Transistors T_{p0}, T_{p1}, and T_{p2} have the same size, and J is the biasing current. Therefore, we get $I_{cm} = \frac{1}{2}\left(I_{d+} + I_{d-}\right)$, which is the common-mode component.

(c) feeding to the following circuit block by wiring the corresponding outputs together. The input currents to the following circuits are $I_{d+} - I_{cm}$ and $I_{d-} - I_{cm}$, rather than I_{d+} and I_{d-}. Therefore, no common-mode components propagate to the following circuits.

Common-mode components are not propagated from one block to another and thus they have no influence on the fully differential signal processing. The penalty of using the CMFF is only the use of current mirrors, eliminating the drawbacks of using CMFBs. We will show in the following sections the use of the CMFF in different fully differential SI circuits.

The drawback of the CMFF shown in figure 3.5 is the power dissipation due to the extra branches needed to process the common-mode currents. Since transistors T_{n2} and T_{n3} have the half size of transistors T_{n0} and T_{n1}, the bias current of T_{p0} is J as well. Therefore, the total current consumption is $3J$. This drawback can be circumvented by using the current scaling technique illustrated in figure 3.6.

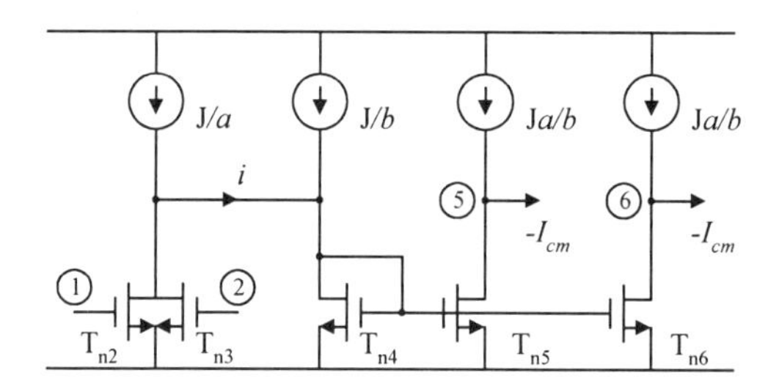

Fig. 3.6. Common-mode current generation with the current scaled.

Shown in figure 3.6 is the common-mode current generation with the possibility of saving power. The dimension ratio of the transistors T_{n2} and T_{n3} is the dimension ratio of the transistors T_{n0} and T_{n1} (shown in figure 3.5 (a)) divided by a factor of $2a$. Therefore, the bias current for transistors T_{n2} and T_{n3} can be reduced to J/a. The intermediate output current i is therefore given by $i = \frac{1}{2a}\left(I_{d+} + I_{d-}\right)$. Since the current i represents the scaled common-mode current, it is usually very small. Only very small bias current (J/b) is needed for the transistor T_{n4}. Transistors T_{n5} and T_{n6} are a times larger than the transistor T_{n4} and therefore their bias currents are a times larger than the bias current of transistor T_{n4}. Due to the current mirror formed by transistors T_{n4}, T_{n5} and T_{n6}, the output current is therefore given by $I_{cm} = a \cdot i = \frac{1}{2}\left(I_{d+} + I_{d-}\right)$ and the correct common-mode current is generated. The total current consumption of the common-mode current generation is $\left(\frac{1}{a} + \frac{2a+1}{b}\right)J$. By properly choosing the coefficients a and b, a power saving by several folds is feasible. Due to the large scaling factors, the accuracy may suffer. However, it really does not introduce any problem in that the purpose of the common-mode control is only to guarantee that common-mode currents do not affect the fully-differential signal processing. Small residue common-mode currents are tolerable for applications such as in A/D converters, because only the difference between the fully differential branches is quantized to generate the digital output.

If transistor T_{n2} and T_{n3} have the half size of the transistors T_{n0} and T_{n1} shown in figure 3.5 (a), transistors T_{n4}, T_{n5}, T_{n6} have the same size, and the coefficients a and b are unity, the common-mode generation circuit shown in figure 3.6 will consume a current of $4J$.

3.5. Fully Differential Class-AB SI Circuits

Saving power is crucial in portable electronics systems. Class-AB circuits offer the potential to realize power efficient SI filters and data converters. However, the serious drawback of class-AB SI circuits is the drastic change in the transconductance with a large input current in respect to the small bias current. The strong dependence of the transconductance on the input current introduces large distortions in SI circuits and therefore class-AB SI circuits are not very good for low-distortion applications.

In figure 3.7, we show a fully differential class-AB SI memory cell [32-34].

Unlike other class-AB memory cells that used circuit techniques to reduce the output conductance [35, 36], this class-AB memory cell uses GGAs to increase the input conductance.

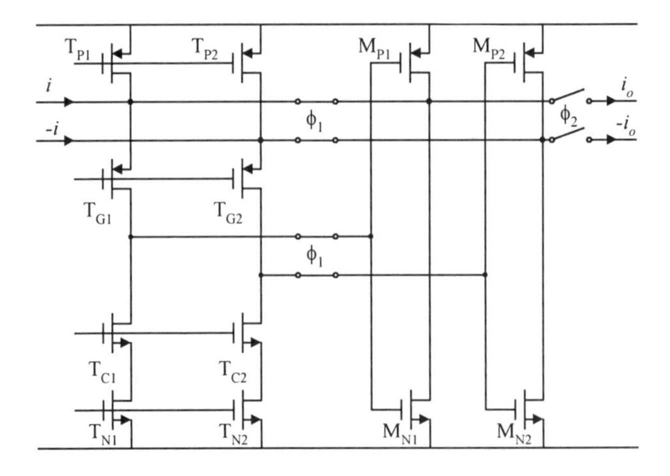

Fig. 3.7. Fully differential class-AB SI memory cell.

The memory cell comprises one pair of memory transistors and one pair of GGAs. The memory transistor pair consist of transistors M_N and M_P, and the GGA consists of the grounded-gate transistor T_G, current biasing transistor T_P and cascode current bias transistors T_C and T_N. The input conductance of the memory cell of figure 3.7 is given by

$$g_{in} = \left(g_{mN} + g_{mP}\right) \cdot A_{GGA} \approx \left(g_{mN} + g_{mP}\right) \cdot \frac{g_{mG}}{g_{dsG}} \tag{3.7}$$

where g_{mN} is the transconductance of the n-type memory transistor M_{N1} (and M_{N2}), g_{mP} is the transconductance of the p-type memory transistor M_{P1} (and M_{P2}), A_{GGA} is the gain of the GGA, g_{mG} is the transconductance of the grounded-gate transistor T_G, and g_{dsG} is the output conductance of the grounded-gate transistor T_G. (The output conductance of the cascode current source for T_G is relatively small and is therefore neglected.) The input conductance can be increased by 20 ~ 40 dB for a practical design. The output conductance of the memory cell shown in figure 3.7 is the summation of the output conductance of the n-type memory transistor M_{N1} (and M_{N2}) and the output conductance of the p-type memory transistor M_{P1} (and M_{P2}). Therefore, the error due to the finite input-output conductance ratio can be reduced substantially (by the gain of the GGAs).

The class AB memory cell bears some resemblance to the class A memory cells [28-30] shown in figure 3.3, where the p-type memory transistors are used as current biasing transistors. The class AB configuration as shown in figure 3.7 allows a more power efficient realization of SI circuits, because the input current can be larger than the quiescent current in the memory transistor that can be designed to be small. However, a large input current with respect to the bias current introduces large distortions due to the large change in the transconductance. For low distortion applications, class-AB SI circuits should be avoided.

The class AB configuration itself reduces the charge injection error if we use an n-type transistor as the switch for the n-type memory transistor and a p-type transistor as the switch for the p-type memory transistor [35]. And the fully differential structure reduces the charge injection error as well [1, 28]. The settling behavior is similar to that of the class A memory cells [28-30] shown in figure 3.3.

To ensure a proper operation, every transistor should be in its saturation region. Therefore, the minimum power supply voltage is given by

$$\begin{aligned} V_{dd} \geq & \left|\left(V_{gs}-V_T\right)_P\right| + \left|\left(V_{gs}-V_T\right)_G\right| + \left(V_{gs}-V_T\right)_C + \left(V_{gs}-V_T\right)_N \\ & + \left(\sqrt{1+m_i}-1\right)\cdot\left|\left(V_{gs}-V_T\right)\right| \end{aligned} \tag{3.8}$$

and

$$V_{dd} \geq \left|\left(V_T\right)_{MP}\right| + \left(V_T\right)_{MN} + \left(\sqrt{1+m_i}\right)\cdot\left|\left(V_{gs}-V_T\right)\right| \tag{3.9}$$

where $\left(V_{gs}-V_T\right)_P, \left(V_{gs}-V_T\right)_G, \left(V_{gs}-V_T\right)_C, \left(V_{gs}-V_T\right)_N$, and $\left(V_{gs}-V_T\right)$ are the quiescent saturation voltages of transistors T_P, T_G, T_C, T_N, and M_N (or M_P), respectively, $\left(V_T\right)_{MP}$ and $\left(V_T\right)_{MN}$ are the threshold voltages of memory transistors M_P and M_N, and m_i is the signal modulation index.

From equations (3.8) and (3.9) it is seen that the use of a low-power supply voltage, say 3.3 V, is possible, given the threshold voltages around 1 V, even with large input currents.

However, it is not shown in figure 3.7 the control of common-mode currents. The CMFF principle discussed in the preceding section can be used to control the common-mode currents. Shown in figure 3.8 is the CMFF circuit for the circuit shown in figure 3.7.

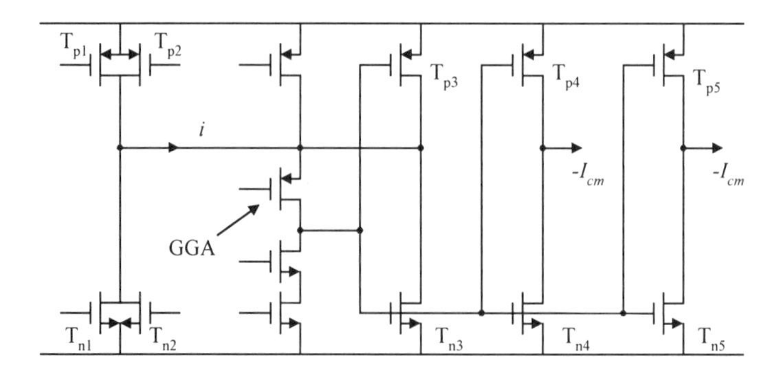

Fig. 3.8. CMFF for the class AB SI circuit shown in Fig. 3.7.

Transistors T_{n1} and T_{n2} have the half size of transistors M_{N1} and M_{N2} in figure 3.7 and transistors T_{p1} and T_{p2} have the half size of transistors M_{P1} and M_{P2} in figure 3.7. The gates of transistors T_{n1} and T_{n2} are connected with the gates of transistors M_{N1} and M_{N2} in figure 3.7, respectively. The gates of transistors Tp1 and T_{p2} are connected with the gates of transistors M_{P1} and M_{P2} in figure 3.7, respectively. Therefore, the current i is the common-mode current. A class-AB current mirror consisting of transistors T_{n3}, T_{n4}, and T_{n5} and transistors T_{p3}, T_{p4}, and T_{np5} is used to duplicate the common-mode current. A GGA is used at the input of the current mirror to reduce the input impedance. Transistors T_{n3}, T_{n4}, and T_{n5} and transistors T_{p3}, T_{p4}, and T_{p5} have the same size, respectively. Since the input current (i.e., the common-mode current) to the current mirror is very small, the quiescent currents for these transistors do not have to be large. The output currents are combined with the output currents of the circuit shown in figure 3.7 to cancel the common-mode current.

By cascading two memory cells and sharing the GGAs, we have a delay line. In figure 3.9, we show the measured power spectrum of the delay line with a single 3.3-V power supply. When the input is a 8-μA 5-kHz sinusoid and the clock frequency is 5 MHz, the measured THD is less than - 50 dBc. (The bias current is about 20 μA.) When the input current further increases, the THD increases significantly due to the strong dependency of the transconductance on the input current of class AB circuits. The presence of the dominating second harmonic is also due to the fact that the transconductance of the differential branches differs.

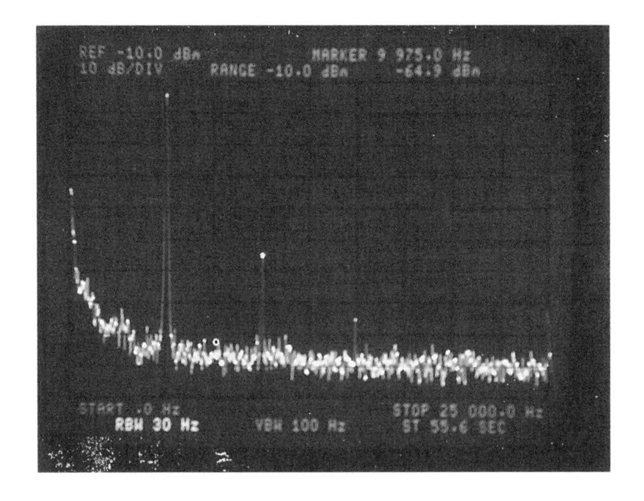

Fig. 3.9. Measured power spectrum of the delay line consisting of two SI memory cells of Fig. 3.7. THD < - 50 dBc.

3.6. HIGH SPEED SI CIRCUITS

For high frequency applications, using GGAs to increase the input conductance is not a good choice due to the explicit loop that limits the speed. Also large transient glitches can be directly coupled onto the gate via the parasitic drain-gate capacitance of the memory cell, increasing the errors. For high speed applications, cascode structures are a better choice. In figure 3.10 we show a fully differential SI memory with cascode transistors.

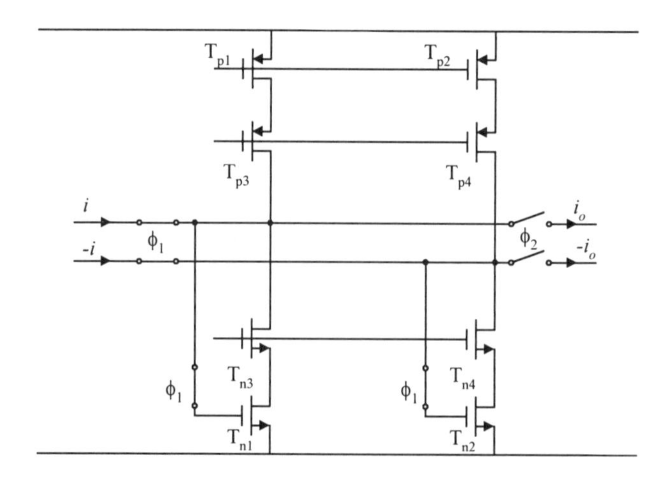

Fig. 3.10. Fully differential SI memory cell with cascodes.

Both the memory transistors T_{n1} and T_{n2} and the bias transistors T_{p1} and T_{p2} have cascodes. The output conductance is given by

$$g_{out} \approx \frac{g_{dsn1}}{g_{mn3}/g_{dsn3}} + \frac{g_{dsp1}}{g_{mp3}/g_{dsp3}} \tag{3.10}$$

where g_{dsn1} is the output conductance of the memory transistor T_{n1} (and T_{n2}), g_{mn3} is the transconductance of the cascode transistor T_{n3} (and T_{n4}), g_{dsn3} is the output conductance of the cascode transistor T_{n3} (and T_{n4}), g_{dsp1} is the output conductance of the bias transistor T_{p1} (and T_{p2}), g_{mp3} is the transconductance of the cascode transistor T_{p3} (and T_{p4}), and g_{dsp3} is the output conductance of the cascode transistor T_{p3} (and T_{p4}).

It is seen that by using cascodes, the output conductance is decreased by the gain of the cascode transistor, while the input conductance is still equal to the transconductance of the memory transistor T_{n1} (and T_{n2}). Since the gain of the cascode transistor is in the range of 20 ~ 40 dB, the error due to the finite input-output conductance ratio is reduced substantially.

Due to the cascode transistor, the influence of the switching glitch is also reduced. The switching glitch at the output node is only directly coupled onto the gate of the cascode transistor, not directly onto the gate of the memory transistor. Since the gate of the cascode transistor is biased at a low impedance, the influence of the glitch is minimized. Notice, due to the limited gain of the cascode transistor at high frequencies, there is still a variation in the drain of the memory transistor which can be coupled onto the gate of the memory cell, introducing distortions.

Duc to the local feedback formed by the cascode transistor, an extra pole is introduced at the source of the cascode transistor. As long as the parasitic pole at the source of transistor T_{n3} (T_{n4}) is much higher than the dominant pole formed by the gate capacitance of transistor T_{n1} (T_{n2}) and its transconductance, a single pole settling can be guaranteed [1]. This can be easily guaranteed by increasing the transconductance of the cascode transistor T_{n3} (T_{n4}) and reducing the parasitic capacitance at the source. By using the cascode transistor, the output conductance is then decreased by the gain of the cascode transistor. Increasing the transconductance of the cascode transistor decreases the output conductance of the memory cell and increase the parasitic pole frequency. Therefore, by cascoding, both high speed and large input-output conductance ratio can be achieved.

However, in the above discussion, the switch-on resistance is neglected. Large switch-on resistance introduces another pole which may not be high enough. This could destroy the possibility of the high speed operation, especially with a reduced supply voltage.

Referring to figure 3.10, if the input current flows into the memory cell and is very large, the gate potential of T_{n1} (or T_{n2}) must settle to be a large value. Suppose the voltage sampling switch (sampling the voltage on the gate of the memory transistor) controlled by ϕ_1 is an NMOS transistor. Its switch-on resistance becomes very large with the settling due to the significantly reduced gate source voltage and the increased threshold voltage (due to the body effect). Therefore, the memory cell needs more time to settle. The settling error is signal dependent. Even when transmission gates are used as voltage sampling switches, the switch-on resistance is still relatively large with a reduced power-supply voltage. One way to reduce the switch-on resistance is to locally increase the clock drive.

In SC circuits, the potential of switched nodes varies from zero to a value close to the power supply voltage. In SI circuits, however, the voltage swings of the switched nodes are quite limited. For example, for the memory cell shown in figure 3.10, its gate potential can never be less than the threshold voltage of the memory transistor. Therefore, to completely cut off the voltage sampling switch, the low clock voltage level only has to be less than the sum of the threshold voltages of the memory cell and the voltage-sampling switch transistor, not necessarily to be zero. This observation leads to the use of level shifting the clock voltage levels as shown in figure 3.11.

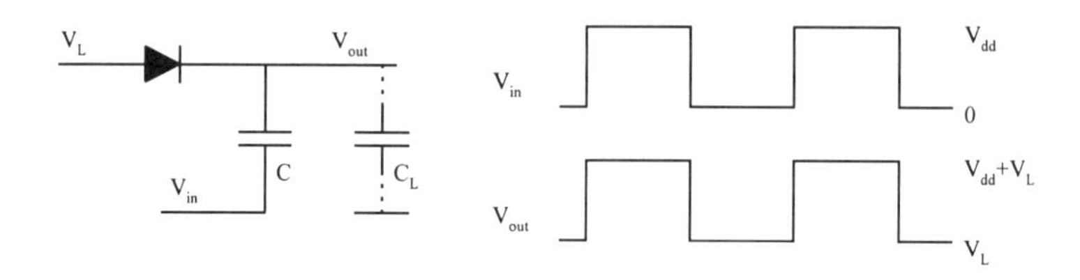

Fig. 3.11. Diode pumping to level-shift clock voltage levels for voltage sampling switches.

In figure 3.11, the diode pumping is used to level-shift the clock voltage levels. The reference voltage V_L is chosen to guarantee a complete cut-off of the voltage-sampling switch.

The benefits of using the scheme rather than doubling the high clock voltage level and keeping the low clock voltage level are 1) simpler circuitry, and 2) reduced clock feedthrough due to the reduced clock voltage swing. By using the diode pumping circuit, only NMOS transistors can be used as switch transistors.

Simulation results are shown in figure 3.12 with a power supply voltage of 3.3 V. The bias current in each branch is about 200 μA. For the memory cell of figure 3.10, transmission gates are used as voltage sampling switches without the level-shifting. When the level-shifting circuit shown in figure 3.11 is used to boost the speed, only NMOS transistors are used as voltage sampling switches and clock voltage levels are level-shifted by 1.2 V. It is seen that the level shifting using the diode pumping effectively boosts the speed.

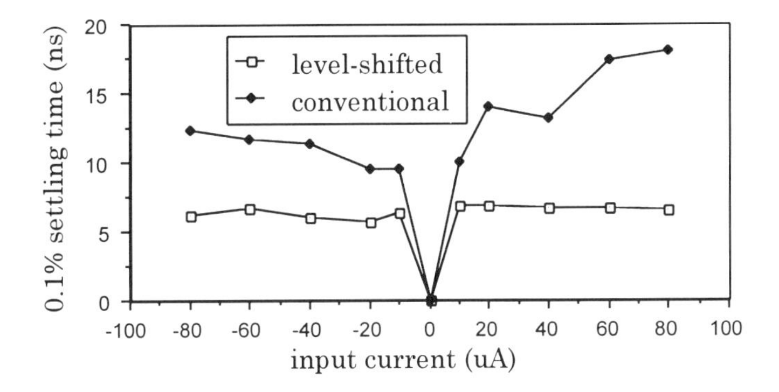

Fig. 3.12. Simulated 0.1% settling time.

To control the common-mode current, the CMFF principle shown in figure 3.5 can be used. Shown in figure 3.13 is the CMFF circuit for the fully differential SI memory cell shown in figure 3.10.

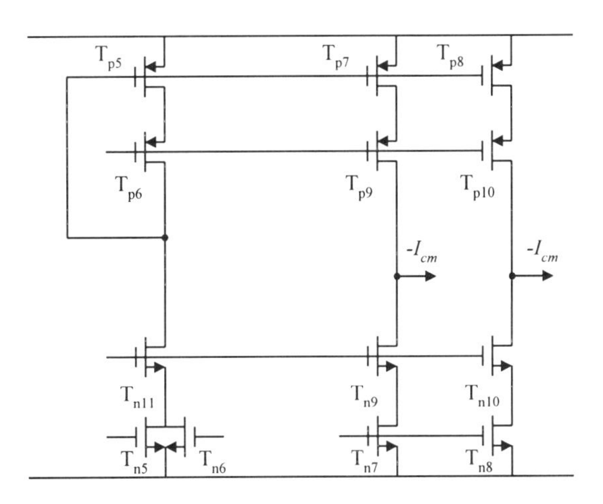

Fig. 3.13. CMFF circuit for the fully differential SI circuit shown in Fig. 3.10.

Transistors T_{n5} and T_{n6} have the half size of transistors T_{n1} and T_{n2} in figure 3.10. The gates of transistor T_{n5} and T_{n6} are connected with the gates of transistors T_{n1} and T_{n2} of figure 3.10, respectively. Transistors T_{p5}, T_{p7} and T_{p8} have the same size. Transistors T_{n7} and T_{n8} have the same size as well. A bias voltage is provided for the gates of matched transistors T_{n7} and T_{n8}. The cascode transistors are also matched. The output current is the common-mode current with the flow direction inverted.

To reduce the power consumption, the current scaled CMFF principle shown in figure 3.6 can be used. In figure 3.14, we show the CMFF circuit with the current scaled. It is based on a specific design.

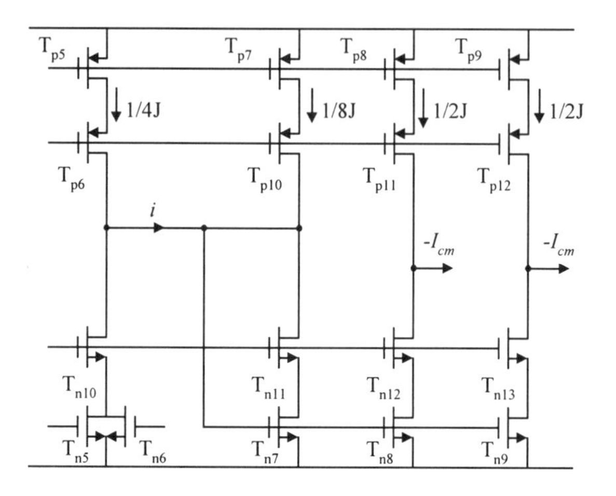

Fig. 3.14. CMFF circuit with the current scaled.

The dimension ratio of transistors T_{n5} and T_{n6} is the 1/8 of the dimension ratio of transistors T_{n1} and T_{n2} in figure 3.10. The bias current of this branch is 1/4 the bias current for transistor T_{n1} (T_{n2}) of figure 3.10. The intermediate output current i is therefore given by $i = \frac{1}{8}\left(I_{d+} + I_{d-}\right)$. The bias current for transistor T_{n7} is 1/8 of the bias current for transistor T_{n1} (T_{n2}). Transistors T_{n8} and T_{n9} are 4 times larger than transistor T_{n7} and their bias currents are four times larger than the bias current for transistor T_{n7}. The cascode transistors (T_{n10}~T_{n13} and T_{p6}, T_{p10}~T_{p11}) are scaled accordingly as well. By doing so, the common-mode current is generated with a current consumption less than half of the current consumption of the CMFF circuit shown in figure 3.13.

In figure 3.15, we show the measured power spectrum of the delay line using the SI memory cell of figure 3.10 with a single 3.3-V supply. The clock voltages are level shifted by 1.2 V. The clock frequency is 100 MHz and the input is a 15-μA 500-kHz sinusoidal. The bias current is about 80 μA. The measured THD is - 47 dBc.

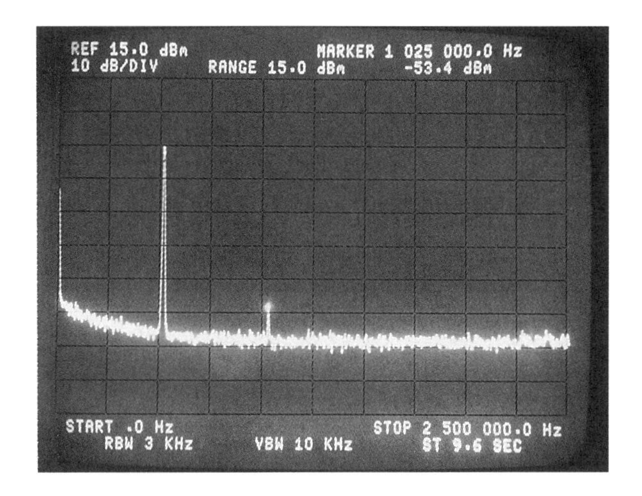

Fig. 3.15. Measured power spectrum of the delay line consisting of the SI memory cells of Fig. 3.10. THD = - 47 dBc.

3.7. ULTRA LOW-VOLTAGE SI CIRCUITS

In SI circuits, the signal carrier is a current not a voltage. Therefore, the signal range is not limited by the power supply voltage as in SC circuits. Another implicit advantage concerning the low-voltage operation is the simplicity of SI circuits. In SC circuits, operational amplifiers are needed that are very difficult to design with a reduced power supply voltage. In SI circuits, we do not need operational amplifiers, and the limit of the power supply voltage is in principle, given by the threshold voltage of one memory transistor. Therefore, the supply voltage can be less than the sum of the threshold voltage values of n and p-type transistors. In figure 3.16, we show the 1.2-V fully differential SI memory cell. The process we use is a 0.8-μ single-poly double-metal digital CMOS process. The threshold voltage of the PMOS transistor is - 0.734 V and that of the NMOS one is 0.844 V.

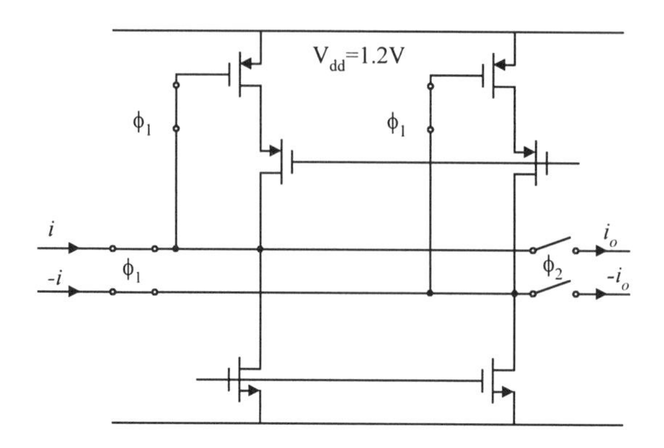

Fig. 3.16. Ultra low-voltage fully differential SI memory cell.

We use PMOS transistors as the memory transistors. To reduce the influence of the drain potential change on the gate voltage due to channel length modulation and more profoundly due to the drain-gate capacitive coupling, we use cascode transistors. NMOS transistors are used to provide bias currents. Long channel devices are used for the memory transistors and bias transistors, and short channel devices are used for the cascode transistors. The transistors are dimensioned to guarantee that every transistor is in its saturation region even with large input currents.

Using PMOS transistors as the memory transistors is advantageous in that they have smaller threshold value in our target process. (Lower flicker noise is not our motivation in that the correlated double-sampling reduces it in the second-generation SI circuits as discussed in Chapter 2.7.) Another advantage is as follows. Since the switched nodes have potentials close to ground, we can use only NMOS transistors as switches and their gate voltages can be higher than the supply voltage V_{dd} without special arrangements of avoiding latch up in the target process (Nwell).

In figure 3.17, we show the simulated transmission error of a single branch of the memory cell. Hspice is used and the transistor model parameters are the normal values of the 0.8-μ single-poly double-metal CMOS process. The transmission error includes all kinds of errors, the settling error, clock feedthrough error, channel length modulation error, capacitive coupling error, and the error in the current source. The clock frequency is 1 MHz and the clock voltages are 0 and 2.4 V.

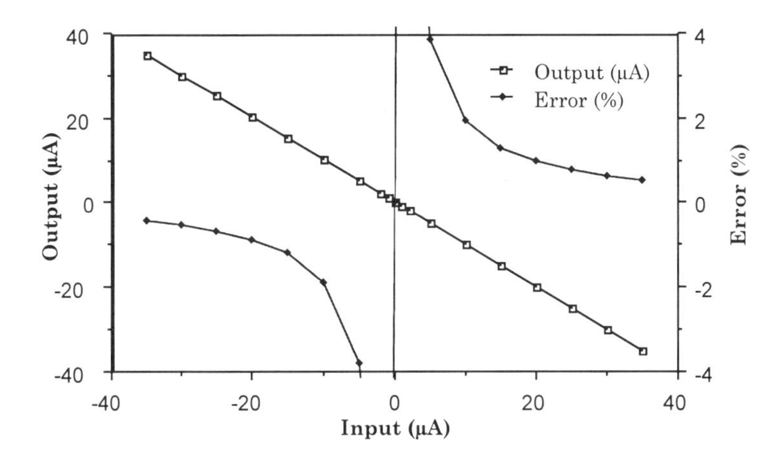

Fig. 3.17. Simulated transmission error of a single branch of the fully differential memory cell of Fig. 3.16. The quiescent current is about 40 μA.

We see from figure 3.17 that the single-ended SI circuit can not deliver an adequate performance. The error is so large that even oversampling A/D converters can not tolerate it.

In figure 3.18, we show the simulation results of the fully differential SI memory cell. We see that the total transmission error is less than 0.036% (-69 dB) even with an input modulation index as large as 87.5%. Therefore, the fully differential SI memory cell can be used in oversampling A/D converters without resorting to the two-step or n-step SI techniques [37, 38], avoiding complex clocking.

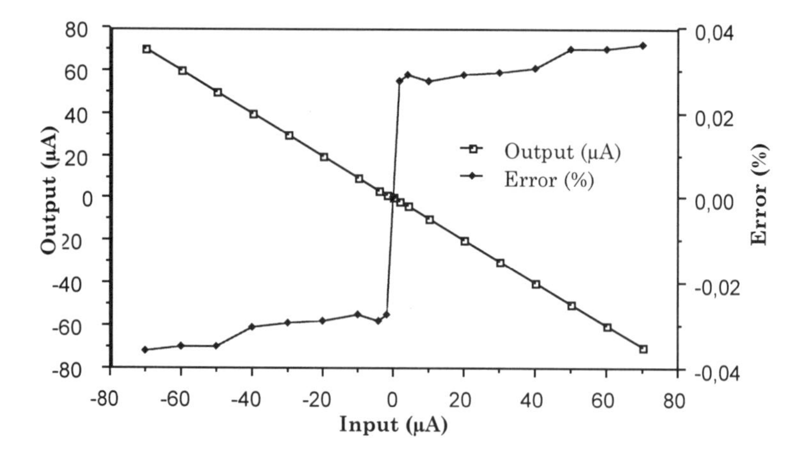

Fig. 3.18. Simulated transmission error of the fully differential memory cell of figure 3.16. The quiescent current of each memory transistor is about 40 µA.

In figure 3.19, we show one realization of the CMFF circuit for the ultra low-voltage fully differential SI circuit of figure 3.16.

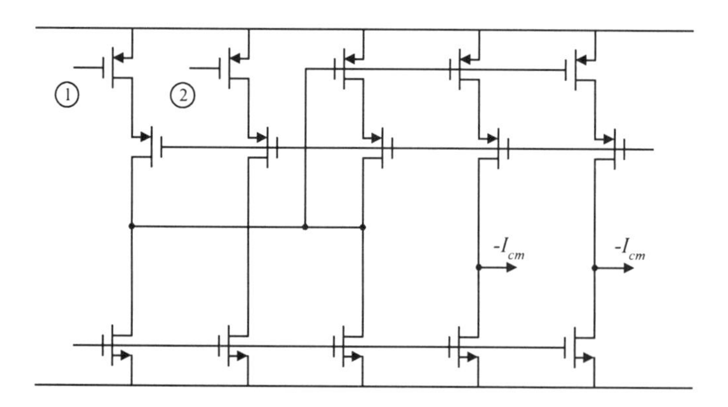

Fig. 3.19. CMFF circuit for the fully differential memory cell of Fig. 3.16.

Nodes 1 and 2 are connected to the gates of the memory transistor pair of figure 3.16, respectively. All the transistors have the half size of corresponding ones in the memory cell. Therefore, the output currents are equal to the common-mode current of the memory cell but with the direction inverted. By connecting the output currents of the memory cell and the output currents of the CMFF circuit together, we generate the output currents that do not contain any common-mode components.

We see that by using the CMFF circuit, we can get rid of the influence of common-mode components on the fully differential signal processing. At the same time, we can guarantee the ultra low-voltage operation due to the fact that the CMFF circuit of figure 3.19 uses the same circuit configuration as the memory cell of figure 3.16.

By cascading two memory cells of figure 3.16, we have a delay line. In figure 3.20, we show the measured power spectrum of the delay line. The supply voltage for the SI delay line is 1.2 V, and the clock voltages are 0 and 3.3 V. When the input is a 24-μA 10-kHz sinusoid and the clock frequency is 1 MHz, the measured THD is less than - 48 dBc. And the bias current for the memory cell is about 40 μA.

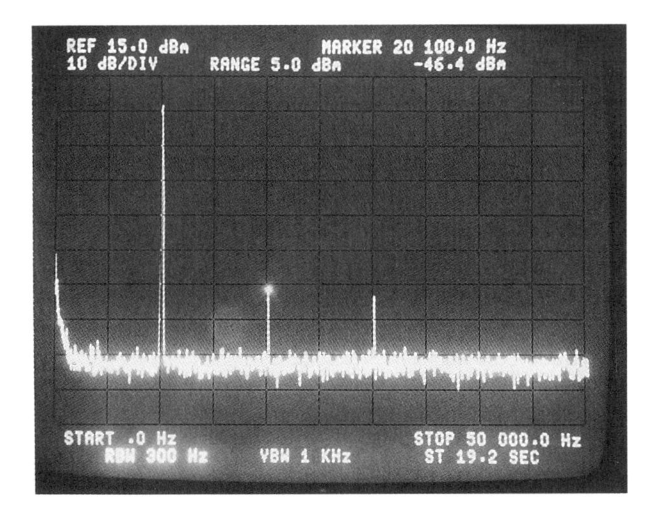

Fig. 3.20. Measured power spectrum of the delay line consisting of the SI memory cells of Fig. 3.16. THD < - 48 dBc.

It is seen from figure 3.20 that SI circuits can operate at a supply voltage less than twice the threshold voltage thanks to the simplicity of SI circuits.

3.8. Two-Step SI Circuits

SI circuits can in principle operate at a very high frequency due to the fact that the basic SI memory cells have a very small RC time constant [28]. However, SI circuits based on basic memory cells deviate from the ideal behavior mainly due to the channel length modulation, drain-gate capacitive coupling, and switch charge injection, and all errors increase with the bandwidth. Different circuit techniques can be used to improve the performance of SI circuits and have been discussed in this chapter. There exists another technique which is called the two-step switched-current (S^2I) technique [37]. This technique provides a total error reduction through an overall circuit operation rather than by the piecemeal application of circuit techniques to suppress individual errors. Therefore, it is expected that S^2I circuits can deliver high performance at high frequencies.

An S^2I memory cell is shown in figure 3.21 (a), an S^2I memory cell with cascodes is shown in figure 3.21 (b), and the clock waveform is shown in figure 3.21 (c).

The input clock phase ϕ_1 is split into two sub clock phases ϕ_{1a} and ϕ_{1b}. During clock phase ϕ_{1a}, the coarse memory transistor M_C sinks the input current. During clock phase ϕ_{1b}, the gate of M_C is open while the input current is kept on so that the fine memory transistor M_F can access the error generated in the coarse memory transistor M_C. Then during clock phase ϕ_2, the currents from the two memory cells are combined to generate the output current i_o, i.e.,

$$i_o = i_C + i_F = (-i + \delta_C) + (-\delta_C + \delta_F) = -i + \delta_F \tag{3.11}$$

where i_C is the output current of the coarse memory cell M_C, i_F is the output current of the fine memory cell M_F, δ_C is the error current generated in the coarse memory cell M_C, δ_F is the error current generated in the fine memory cell M_F, and i is the input current.

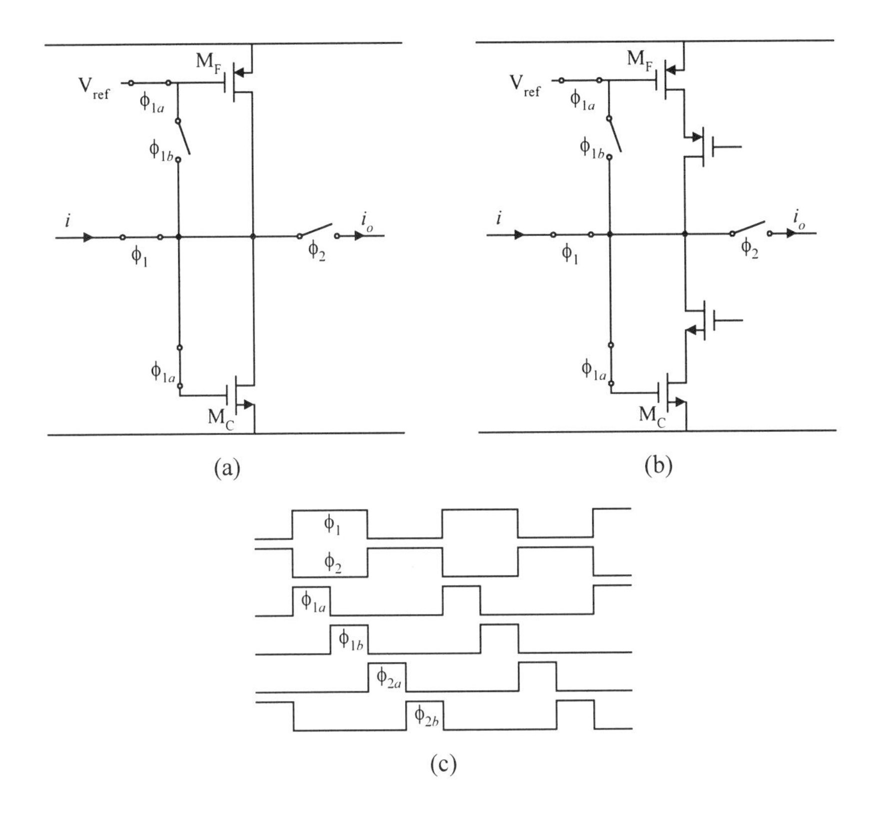

Fig. 3.21. (a) S^{2}I memory cell, (b) cascode S^{2}I memory cell, and (c) clock waveform for the S^{2}I technique.

Since the input to the fine memory cell is only the error current generated in the coarse memory cell, the signal transmission error ε is

$$\varepsilon = \varepsilon_C \cdot \varepsilon_F \tag{3.12}$$

where ε_C and ε_F are the combined errors of the coarse and fine memory cell, respectively.

The high performance results from the fact that during the input phase ϕ_{1b} and output phase ϕ_{2b}, the drain potential of the memory transistors are the same, akin to a 'virtual ground'. However, with a reduced power supply voltage, say 3.3 V, the voltage drop across the current-steering switches controlled by ϕ_1 and ϕ_2 are rather large when a large input current is present. Therefore, the drain potential differs during the input phase ϕ_{1b} and output phase ϕ_{2b}. This drain potential change has a significant influence on the performance due to the channel length modulation and capacitive coupling. For high bandwidth circuits, short channel devices are used and the channel length modulation is significant. The drain-gate capacitance in short channel devices is not much smaller than the gate-source capacitance. The drain potential change is capacitively coupled onto the gate voltage which introduces a significant error. Also noted is that the voltage drop across the current-steering switches is signal dependent and therefore this kind of errors introduce distortions in S^2I circuits.

Reducing the voltage drop across the current-steering switches is the way to reduce the errors. This can be accomplished by using very large switch transistors and/or increasing the clock voltages. The adverse effect is large transient current spikes which degrade the performance [31]. A better way is to use the cascode technique to reduce the effects of the channel length modulation and the capacitive coupling. The cascoding has little effect on the speed and low-voltage operation as long as the saturation voltage of the cascode transistor is designed very small and the parasitic pole frequency is high [1].

In figure 3.22, we show the simulation results of fully differential S^2I memory cells. The current steering switches are transmission gates with W/L = 16/0.8 and the voltage sampling switches are minimum-sized NMOS transistors (W/L = 2/0.8). The biasing current in each branch is about 110 μA, the power supply voltage is 3.3 V, and the clock frequency is 1 MHz. The output is fed to an identical memory cell and the value is measured during output phase ϕ_{2b}.

It is seen that cascoding significantly reduces the transmission error (less than 0.05%). Note that in some applications using delay elements as the building blocks (e.g., FIR filters), the influence of the voltage drop across the current-steering switches in the successive memory cells tends to reduce each other.

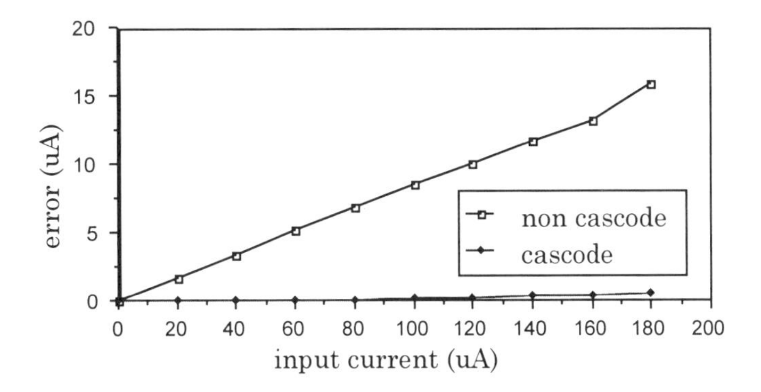

Fig. 3.22. Simulated transmission errors of S^{2}I memory cells with and without cascode transistors.

The CMFF technique can also be applied to S^{2}I circuits. In figure 3.23 we show a CMFF circuit for the S^{2}I memory cell shown in figure 3.21 (b).

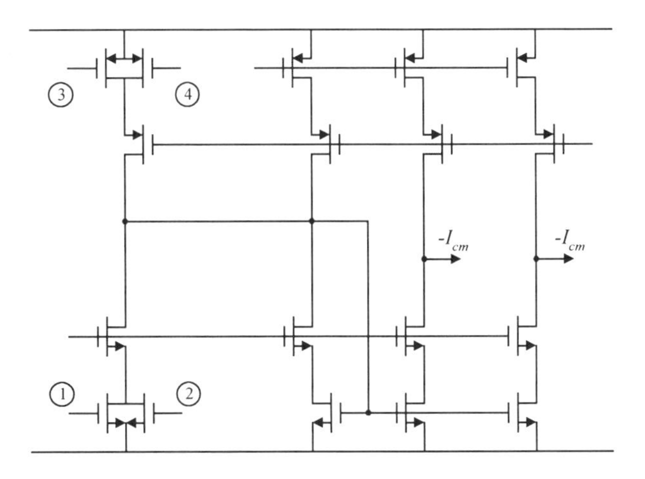

Fig. 3.23. CMFF circuit for the S^{2}I circuit of Fig. 3.21 (b).

Nodes 1 and 2 are connected to the gates of the coarse memory transistor pair and nodes 3 and 4 are connected to the gates of the fine memory transistor pair, respectively. All the transistors have half size of the corresponding transistors shown in figure 3.21 (b). The output current I_{cm} is equal to the common-mode current of the memory cell that the CMFF circuit is connected to. The two output currents of the CMFF circuit are then connected to the differential outputs of the memory cell during its output clock phase.

S^2I memory cells have the same speed performance as basic SI memory cells. In estimating the speed performance, the switch-on resistance of the voltage-sampling switch is neglected [28]. However, with a reduced power supply voltage, the assumption does not hold if the input current modulation index is large. To operate at high speed, the level shifter of the clock drive shown in figure 3.11 is necessary.

In figure 3.24, we show the measured power spectrum of the delay line using the S^2I memory cell of figure 3.21 (b) with a single 3.3-V supply. The clock voltages are level shifted by 1.2 V. The effective clock frequency is 20 MHz and the input is a 15-μA 20-kHz sinusoid. The measured THD is - 50 dBc.

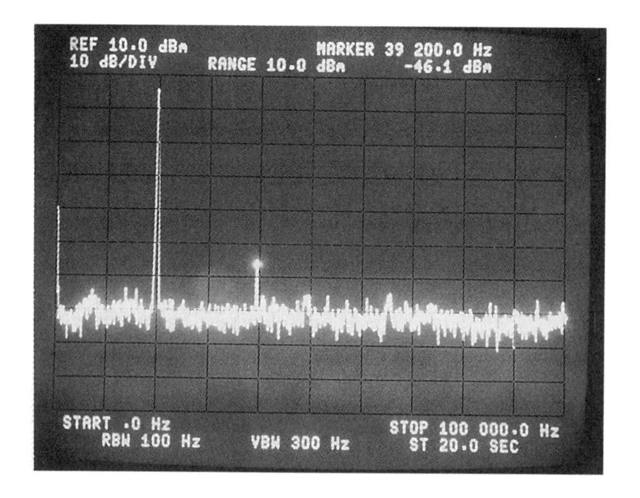

Fig. 3.24. Measured power spectrum of the delay line consisting of the S^2I memory cells of Fig. 3.21 (b). THD = - 50 dBc.

The delay line using the S^2I memory cells of figure 3.21 (b) gives similar performances compared with the SI memory cells of figure 3.10. The main reason is that the harmonic distortion due to the transient current spikes dominates in the second-generation SI circuits [31]. Also the variation of the transconductance depending on the input current introduces distortion, though the S^2I technique effectively reduces the clock feedthrough error and settling error [37].

3.9. SI Technique in BiCMOS

The ultimate performance of SI circuits is primarily determined by the transconductance g_m of an MOS transistor and the capacitance C_g seen by its gate. Although high speed operation (around 100 MHz) is possible as discussed in Chapter 3.6, the performance of speed and accuracy is limited by the technique itself. In order to have a high accuracy, a large gate capacitance C_g is usually required, since the circuit errors are inversely proportional to C_g. Therefore, increasing g_m is the only way to increase the speed. With the same bias current, the transconductance of an MOS transistor is considerably lower than that of a bipolar transistor. Therefore, utilization of bipolar transistors could increase the speed and/or improve the accuracy. The BiCMOS technology opens the possibility of using both MOS and bipolar transistors.

There reported a technique in [39] for high speed sampled-data signal processing in the BiCMOS technology. It breaks the limitation of g_m/C_g of SI circuits by utilizing bipolar transistors. It first converts the current to a voltage by a transresistor and then converts the voltage to a current by a transconductor. The voltage is sampled and held at the input of the transconductor whose input device is an MOS transistor. However, the conversion accuracy is determined by the absolute value of components. For example, the resistor determines the transresistance value, and the transistor size and operation condition determine the transconductance value. (Notice that the changes in these values do not necessarily introduce harmonic distortions as long as the conversions are linear.) Therefore, the technique is sensitive to process variation. The other drawback is its complexity.

In figure 3.25, we show another technique utilizing a BiCMOS process for SI circuits. The technique utilizes a composite transistor consisting of an MOS transistor M0 and a bipolar transistor Q0. The MOS transistor M0 is in common-drain configuration and the bipolar transistors Q0 and Q1 are in common-emitter configuration as shown in figure 3.25. Current source J0, I0, and I1 provide bias currents for transistors M0, Q0, and Q1, respectively. Capacitor C_0 represents all the capacitance at the gate of transistor M0, and C_1 represents all the capacitance at the source of transistor M0.

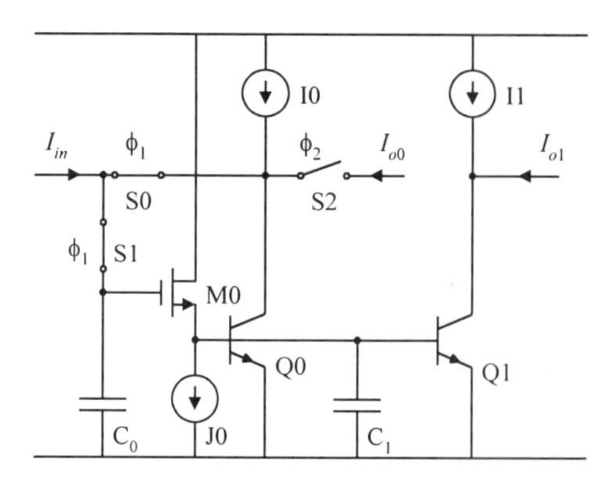

Fig. 3.25. Circuit configuration of a second-generation SI memory cell in BiCMOS.

During clock phase ϕ_1, switches S0 and S1 are closed, and S2 is open. The input current I_{in} flows into the collector of transistor Q0 and makes the base-emitter voltage change correspondingly. Due to the common-drain configuration of transistor M0 (its gate-source voltage does not change), the potential at the gate of transistor M0 changes proportionally as well. When the stable condition is reached, a potential at the gate of transistors M0 is created to change the base-emitter voltage of transistor Q0 to sink (or source) the input current into transistor Q0. Since transistors Q0 and Q1 have the same base-emitter voltage, the output current I_{o1} is equal to the input current I_{in}, if they have the same emitter area.

During clock phase ϕ_2, switches S0 and S1 are open, and S2 is closed. The gate of the MOS transistor M0 is isolated and the potential at the gate is held. Since the gate-source voltage of transistor M0 is constant, the base-emitter voltage of transistor Q0 does not change. Therefore, the collector current of Q0 does not change. The output current I_{o0} is equal to the input current I_{in} that was the input to transistor Q0 during clock phase ϕ_1. Since transistors Q0 and Q1 have the same base-emitter voltage, the output current I_{o1} is equal to the output current I_{o0}, if they have the same emitter area.

Therefore, the output current I_{o0} is the memory of the input current I_{in} and the output current I_{o1} realizes a track-and-hold function performed on the input current I_{in}. Since same devices (M0 and Q0) are used both as input and as output devices, there is no mismatch between the input current I_{in} and the output current I_{o0}, just as in a second-generation SI memory cell [1]. A scaling factor between the output current I_{o1} and the input current I_{in} can be realized by choosing a different emitter area for transistor Q1.

The speed of the circuit is determined by the settling time when switches S0 and S1 are closed. Neglecting the switch-on resistance of the switch transistors, the system is a two-pole system. The dominant pole frequency ω_0 is given by

$$\omega_0 = \frac{g_{mQ0}}{C_0} \tag{3.13}$$

where g_{mQ0} is the transconductance of the bipolar transistor Q0 and C_0 is the total capacitance at the gate of M0. The non-dominant pole frequency ω_n is given by

$$\omega_n = \frac{g_{mM0}}{C_1} \tag{3.14}$$

where g_{mM0} is the transconductance of the MOS transistor M0 and C_1 is the total capacitance at the source of transistor M0.

For SI circuits in CMOS process, the dominant pole frequency is determined by the total capacitance seen by the gate of an MOS transistor and the transconductance of the MOS transistor. Due to the higher transconductance of bipolar transistors, the SI technique in BiCMOS has a higher speed performance if the non-dominant frequency is sufficiently high.

This can be satisfied in the circuit design by minimizing the capacitance at the source of M0, especially when a reasonably large capacitance C_0 is used to reduce the circuit errors.

We can also trade the speed for the accuracy by using a large capacitance C_0, since the circuit errors are inversely proportional to C_0. One of the error sources in SI circuits in CMOS is due to the drain-gate parasitic capacitance. When the drain potential changes, it couples onto the gate, which introduces excessive errors [31], especially in high speed SI circuits. In the circuit shown in figure 3.25, the drain potential of the MOS transistor is tied to the power supply, therefore, during the switching, the gate voltage is not influenced. The transient glitches can only be coupled onto the base of the bipolar transistor Q0 via the parasitic base-collector capacitance. Since the base of transistor Q0 has low impedance due to the common-drain configuration of the MOS transistor M0, the transient glitches coupled onto the base is damped significantly. The problems associated with the drain-gate parasitic coupling in CMOS SI circuits (see Chapter 2.6) disappear. Therefore, the SI circuits in BiCMOS have smaller errors, both the signal dependent and signal independent errors.

Devices M0 and Q0 are used both as input and as output devices as in the second-generation SI memory cells in CMOS process, mismatch does not introduce any error. However, in most cases we need current mirrors to realize different coefficients as in the case of using transistor Q1, and the mismatch plays an important role. Since the bipolar transistor matching is better than the MOS transistor matching, BiCMOS SI circuits usually have a higher accuracy than CMOS SI circuits.

Finally, it is worthwhile to note the simplicity. Since bipolar transistors have a larger early voltage and a higher transconductance, the input-output conductance ratio of the BiCMOS SI memory cell is high. Therefore, the circuit illustrated in figure 3.25 can function well without a further elaboration. In principle, SI circuits are simple in CMOS too. However, to deal with different errors, e.g., the clock feedthrough errors, finite input-output conductance ratio errors, and the errors due to the gate-drain parasitic coupling, relatively complex circuits and/or clocking are needed, especially for high speed applications (see Chapter II). Also noted is that the technique illustrated in figure 3.25 does not require linear capacitors either, just as the SI technique in CMOS.

Compared to the technique presented in [39], the SI technique in BiCMOS illustrated in figure 3.25 does not demand matching between the transresistor and transconductor, and the circuit scheme is much simpler.

An alternative circuit realization is shown in figure 3.26. It bears resemblance to the first-generation SI memory cell in CMOS [1]. In figure 3.26, we use different devices as the input and output devices. Transistors M0 and Q0 are used as the input devices, and transistors M1 and Q1 are used as the output devices. Current sources provide bias currents for them.

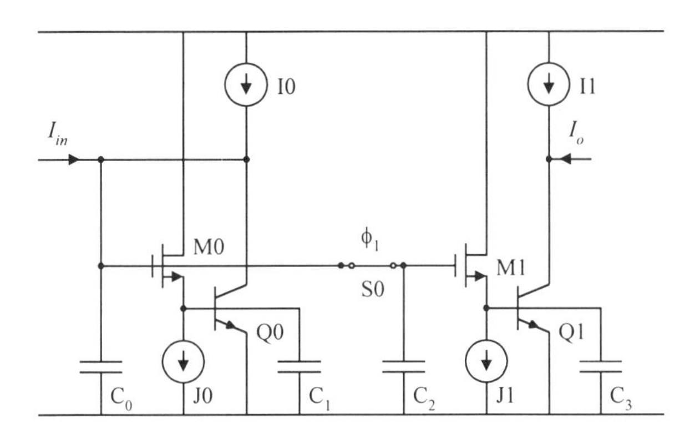

Fig. 3.26. Circuit configuration of a first-generation SI memory cell in BiCMOS.

Suppose that transistors M0 and M1 have the same size, and transistors Q0 and Q1 have the same size. During clock phase ϕ_1 when switch S0 is closed, the gate voltage of M1 is equal to that of M0 and therefore the base-emitter voltages of transistors Q0 and Q1 are equal. This makes the collector currents of Q0 and Q1 equal. Therefore the output current I_o is equal to the input current I_{in}. During clock phase ϕ_2 when switch S0 is open, the gate of M1 is isolated and the potential is held. This makes the base-emitter voltage of Q1 constant and therefore the collector current constant. The output current I_o is held constant. Therefore, the circuit realizes a track-and-hold function just as the first-generation SI memory cell in CMOS. Due to the use of bipolar transistors in the BiCMOS SI circuit shown in figure 3.26, the circuit of figure 3.26 has a better performance than its CMOS counterpart just as outlined above.

For a fully differential realization, we can easily extend the CMFF principle discussed in Chapter 3.4 to the circuits shown in figures 3.25 and 3.26. An example is shown in figure 3.27.

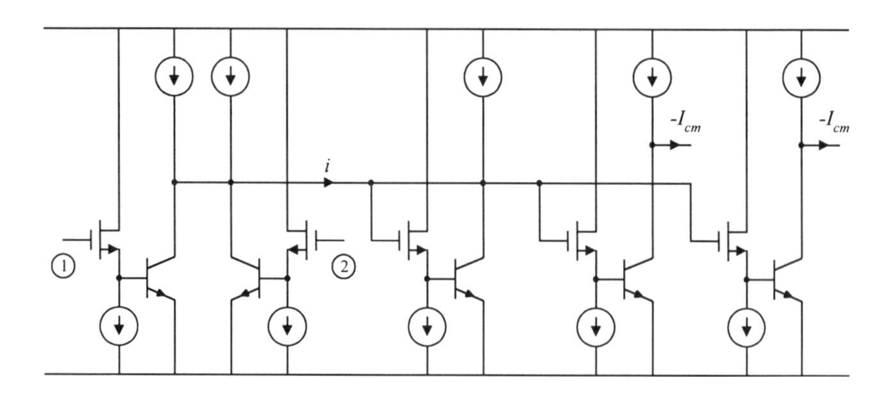

Fig. 3.27. CMFF circuit for the BiCMOS SI circuits.

Nodes 1 and 2 are connected with the gates of the memory transistors of the BiCMOS SI circuits of figures 3.25 or 3.26, respectively. The sizes of all the transistors and bias currents are half of the corresponding parts in the main BiCMOS SI circuits. The generated common-mode current i is then fed to the two-output BiCMOS current mirror with a unity gain to generate the inverted common-mode current. And of course, different current mirrors can be used and the currents can be scaled to save the power as well.

In figures 3.28 and 3.29, we show the simulation results of the circuit of figure 3.25 by using the parameters of a 3.3-V digital BiCMOS process. (The supply voltage is 3.3 V.) We show the input and output currents in figure 3.28. The input current is a 20-MHz 100-μA sinusoid and the clock frequency is 100 MHz. It is evident that a track-and-hold function is realized.

In Fig. 3.29, we show the simulated current errors versus the input currents of a fully differential design based on the circuit scheme shown in figure 3.25. The bias current in each branch is about 360 μA. It is seen that when the sampling frequency is 100 MHz, the error is less 0.55% and the variation is small. This indicates a good linearity. When the clock frequency increases to 250 MHz, the error increases due to the settling error. The error variation is still small when the input current is less than 50% of the bias current, indicating a good linearity.

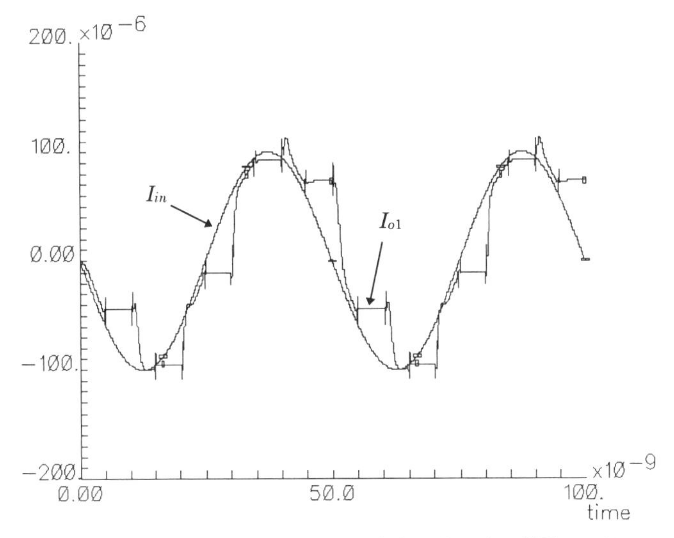

Fig. 3.28. Simulated response of the circuit of Fig. 3.25.

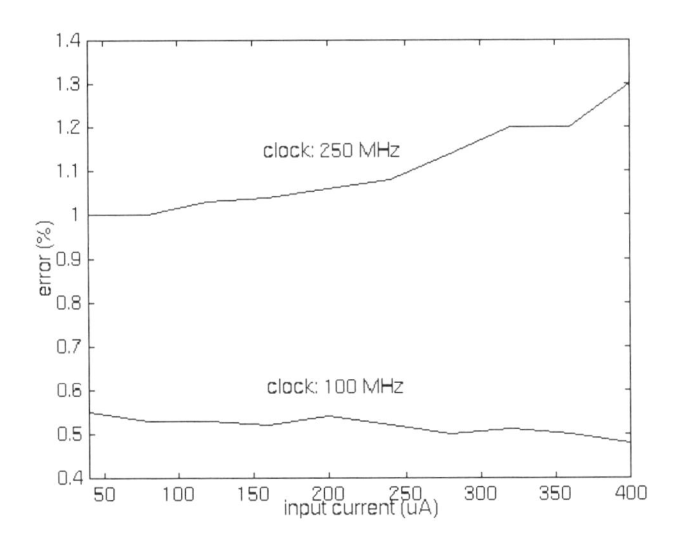

Fig. 3.29. Simulated transmission error of the circuit of Fig. 3.25.

3.10. Other Existing SI Techniques

In this part, we will outline other feasible techniques to improve the performance of basic SI circuits that we have not covered so far. More detailed discussions can be found, e.g., in [1].

A. Operational-amplifier active memory cell

To increase the input-output conductance ratio of a basic SI memory cell, we can use a negative feedback to increase the input conductance. One way is to use a GGA (see Chapters 3.3 and 3.5) and another way is to use an operational amplifier. In figure 3.30, we show the memory cell employing an operational amplifier to increase the input conductance.

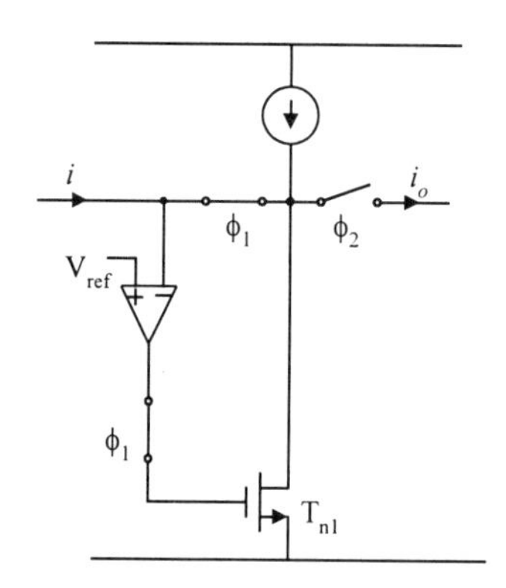

Fig. 3.30. Operational-amplifier active memory cell.

Due to the use of the operational amplifier, the drain potential of transistor T_{n1} is fixed at a reference voltage V_{ref}. The input conductance is given by

$$g_{in} = g_m \cdot A_0 \tag{3.15}$$

where g_m is the transconductance of the memory transistor T_{n1} and A_0 is the DC gain of the operational amplifier.

The input conductance is increased by the DC gain of the operational amplifier and the output conductance is the output conductance of the memory transistor. The input-output conductance ratio is therefore increased by the DC gain of the operational amplifier.

The major drawback of this configuration is the difficulty in achieving a monotonic settling. It is not suitable for high speed applications. Also due to the use of the operational amplifier, the complexity is relatively high.

B. Regulated cascode memory cell

To increase the input-output conductance ratio of a basic SI memory cell, we can also decrease the output conductance. This can be simply achieved by using cascode transistors. The output conductance is then reduced by the gain of the cascode transistor without much speed penalty as we have discussed in Chapter 3.6. If the gain of one cascode transistor does not suffice in some applications, regulated cascode transistors can be used. In figure 3.31 we show the regulated cascode memory cell.

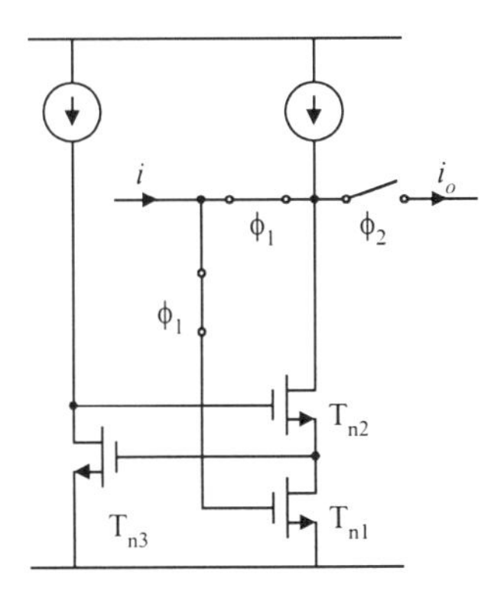

Fig. 3.31. Regulated cascode memory cell.

The only difference between a regulated cascode memory cell and a simple cascode memory cell is the use of the regulating transistor T_{n3}. Due to the local feedback formed by the cascode transistor T_{n2} and the regulating transistor T_{n3}, the output conductance is reduced by the product of the gain of transistor T_{n2} and the gain of transistor T_{n3} and is given by

$$g_{out} \approx \frac{g_{dsn1}}{A_{n2} \cdot A_{n3}} = \frac{g_{dsn1}}{\left(g_{mn2}/g_{dsn2}\right)\cdot\left(g_{mn3}/g_{dsn3}\right)} \tag{3.16}$$

where g_{dsn1} is the output conductance of the memory transistor T_{n1}, A_{n2} is the gain of the cascode transistor T_{n2}, A_{n3} is the gain of the regulating transistor T_{n3}, g_{mn2} is the transconductance of the cascode transistor T_{n2}, g_{dsn2} is the output conductance of the cascode transistor T_{n2}, g_{mn3} is the transconductance of the regulating transistor T_{n3}, and g_{dsn3} is the output conductance of the regulating transistor T_{n3}. Notice that the output conductance of the current source for the memory cell is not considered here. The current source usually takes the form of the regulated cascode configuration and therefore its output conductance is in the same range as the regulated cascode memory cell.

As with the operational-amplifier active memory cell, the main problem is the poor settling behavior due to the extra poles introduced by the feedback loop. Therefore, it is not suitable for high speed applications.

C. Folded-cascode memory cell

The folded-cascode memory cell is shown in figure 3.32. The cascode transistor T_c has the opposite polarity (p type) of the memory transistor T_{n1} (n type). Unlike cascode memory cells, both the memory transistor T_{n1} and the cascode transistor T_c need bias currents.

The input conductance is the transconductance of the memory transistor T_{n1}. The output conductance of the folded-cascode memory cell is decreased by the gain of the cascode transistor Tc, and is given by

$$g_{out} \approx \frac{g_{dsn1}}{A_c} = \frac{g_{dsn1}}{g_{mc}/g_{dsc}} \tag{3.17}$$

where g_{dsn1} is the output conductance of the memory transistor T_{n1}, A_c is the gain of the folded-cascode transistor T_c, g_{mc} is the transconductance of the folded-cascode transistor T_c, and g_{dsc} is the output conductance of the folded-cascode transistor T_c. Notice that the output conductance of the current source for the memory cell is not considered here. The current source usually takes the form of the folded-cascode configuration as well and therefore its output conductance is in the same range as the folded-cascode memory cell.

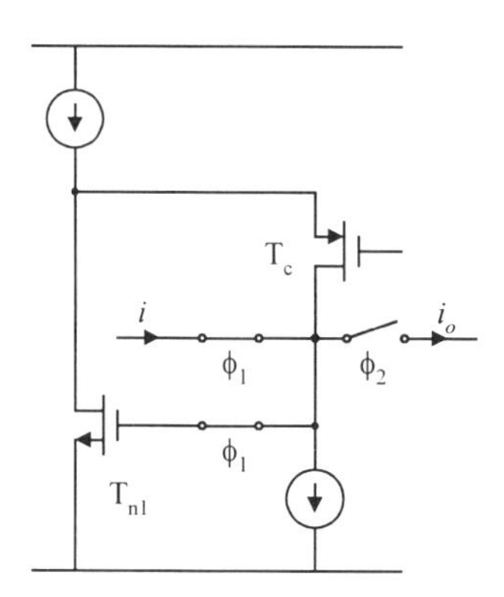

Fig. 3.32. Folded-cascode memory cell.

A monotonic settling is possible if the parasitic pole frequency at the source of the cascode transistor T_c is very high. The behavior of the folded-cascode memory cell is very much the same as the ordinary cascode memory cell as shown in figure 3.10. However, when we cascade two cascode memory cells, the potential at the output node of the first cascode memory cell is set by the gate potential of the second cascode memory cell. With large input currents, this may force some transistors in the first memory cell out of saturation region, increasing the transmission error. This problem is not present in the folded-cascode memory cell [1] shown in figure 3.32. The disadvantage of the folded-cascode memory cell compared with the ordinary cascode memory cell is the extra bias needed, increasing both the power dissipation and the noise.

D. Regulated folded-cascode memory cell

To further decrease the output conductance, the regulated architecture can be applied to the folded-cascode memory cell. In figure 3.33, we show the regulated folded-cascode memory cell. Regulating transistor T_r and its associated bias current are added.

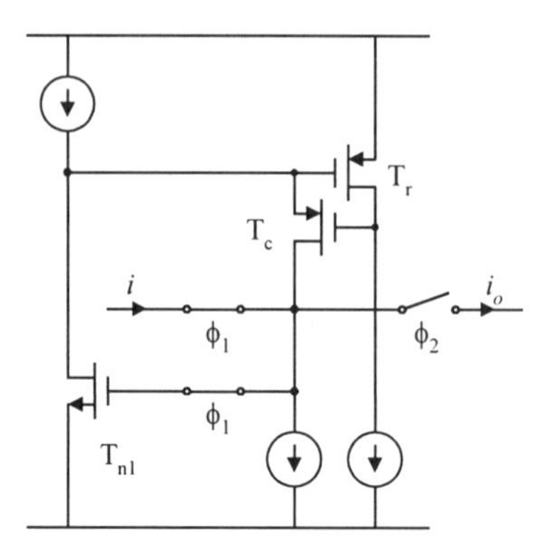

Fig. 3.33. Regulated folded-cascode memory cell.

The input conductance is the transconductance of the memory transistor T_{n1}, and the output conductance is decreased by the product of the gain of the cascode memory transistor T_c and the gain of the regulating transistor T_r and is given by

$$g_{out} \approx \frac{g_{dsn1}}{A_c \cdot A_r} = \frac{g_{dsn1}}{\left(g_{mc}/g_{dsc}\right)\cdot\left(g_{mr}/g_{dsr}\right)} \tag{3.18}$$

where g_{dsn1} is the output conductance of the memory transistor T_{n1}, A_c is the gain of the cascode transistor T_c, A_r is the gain of the regulating transistor Tr, g_{mc} is the transconductance of the cascode transistor T_c, g_{dsc} is the output conductance of the cascode transistor T_c, g_{mr} is the transconductance of the regulating transistor T_r, and g_{dsr} is the output conductance of the regulating transistor T_r. Notice that the output conductance of the current source for the memory cell is not considered here. The current source usually takes the form of the regulated folded-cascode configuration and therefore its output conductance is in the same range as the regulated folded-cascode memory cell.

The regulated folded-cascode memory cell has a behavior similar to the regulated cascode memory cell shown in figure 3.31. (Therefore it is not suitable for high speed operations.) It has an advantage. When we cascade two memory cells, the potential of the output node of the first memory cell set by the gate potential of the second memory cell does not impose a limitation on the saturation voltages in the first memory cell due to the use of the folded-cascode technique [1].

E. Miller-enhanced SI circuit

The well-known Miller effect can be utilized in SI circuits to reduce chip area [40]. A Miller enhanced SI memory cell is shown in figure 3.34.

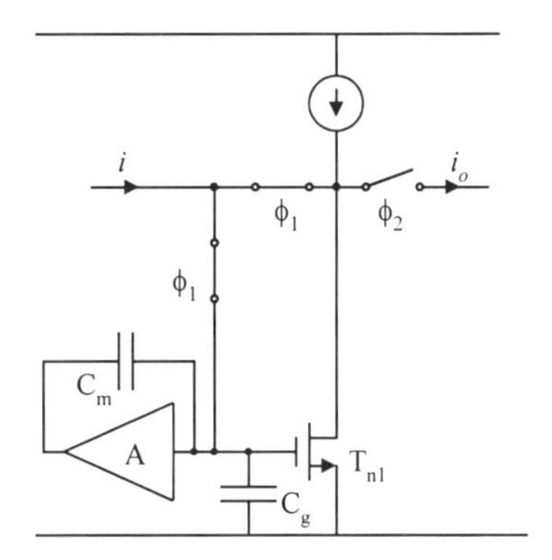

Fig. 3.34. Miller-enhanced SI memory cell.

Due to the Miller effect, the equivalent capacitance at the input of an amplifier is equal to the product of the gain of the amplifier and the capacitance connecting the input and output. The total equivalent capacitance at the gate of the memory transistor T_{n1} is therefore given by

$$C_{in} = (1 + A) \cdot C_m + C_g \tag{3.19}$$

where A is the gain of the amplifier, C_m is the Miller capacitance, and C_g is the total physical capacitance at the gate of the memory transistor T_{n1}.

Since the clock feedthrough error and thermal noise power are inversely proportional to the gate capacitance, increasing the total capacitance can increase the accuracy of SI circuits at the cost of speed. It is seen from equation (3.19), it may be less area consuming by using a Miller capacitor and an amplifier than by using a physical gate capacitor directly, depending on the design of the amplifier. Since the impedance seen at the gate of the memory transistor changes dramatically during input and output phases, care must be taken to stabilize the amplifier [40]. If the gain of the amplifier changes significantly, the error in the memory cell will increase dramatically due to the change of the equivalent gate capacitance.

F. Transconductor-based SI circuit

As discussed in Chapter II, the strong dependency of the transconductance of an MOS transistor on the input current introduces large distortions. If we make the transconductance independent of the input current, then the distortions in SI circuits will decrease significantly. This can be accomplished by replacing the MOS transistor with a transconductor whose transconductance is relatively constant.

Shown in figure 3.35 is a transconductor-based fully differential second-generation SI memory cell [42]. The common-mode control is embedded within the transconductor. The transconductor-based SI memory cell is same as the basic SI memory cell shown in figure 1.1 (b) except that the MOS transistor is replaced by a transconductor. Therefore, the operation principle is the same. Due to the parasitic pole frequencies within the transconductor, it usually requires more capacitance at the input to avoid instability so that the dominant pole frequency due to the transconductance and the total capacitance at the input is much smaller than the parasitic pole frequencies within the transconductor. The extra capacitance can be formed by MOS transistors as shown in figure 3.35. Due to the relatively constant transconductance of the transconductor, distortions decrease significantly.

The transconductor can take a form of an operational transconductance amplifier or any other forms. The key issue in designing the transconductor is the constant transconductance.

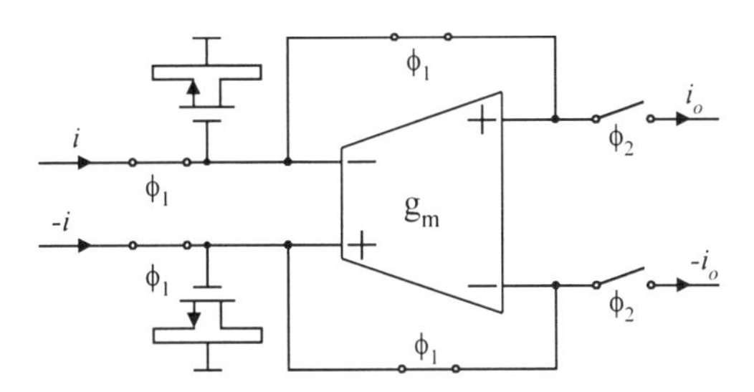

Fig. 3.35. Transconductor-based second-generation SI memory cell.

The advantage of transconductor-based SI circuits is due to the use of transconductors and the serious disadvantage is also due to the use of the transconductors. To have a fast settling behavior, the parasitic pole frequencies in the transconductors must be much higher than the pole frequency due to the transconductance and the memory capacitance. Therefore transconductor-based SI circuits are not so good for high speed applications, though they can have very low distortions. Also due to the use of the transconductors, the circuit complexities increase, making it more difficult to operate with a reduced supply voltage.

G. N-step SI circuit

The n-step SI (S^nI) circuits can be used to achieve an even higher performance than S^2I circuits [38]. The S^nI technique is based on the idea of the S^2I technique. In stead of using two steps, the S^nI circuits use n steps to successively cancel the error introduced by the preceding step. In figure 3.36, we shown the S^nI memory cell where $n = 3$.

The input clock phase ϕ_1 is split into three sub clock phases ϕ_{1a}, ϕ_{1b}, and ϕ_{1c}. During clock phase ϕ_{1a}, the coarse memory transistor T_{n1} sinks the input current i. During clock phase ϕ_{1b}, the gate of T_{n1} is open while the input current i is kept on so that the first fine memory transistor T_{n2} can access the error generated in the coarse memory transistor T_{n1}. During clock phase ϕ_{1c}, the gate of the first fine memory transistor T_{n2} is isolated while the error from the coarse memory transistor T_{n1} is still kept on so that the error generated by the first fine memory transistor T_{n2} is accessed by the second fine memory transistor T_{n3}. Then during output clock phase ϕ_2, all the currents from the different stages are combined to generate the output current i_o.

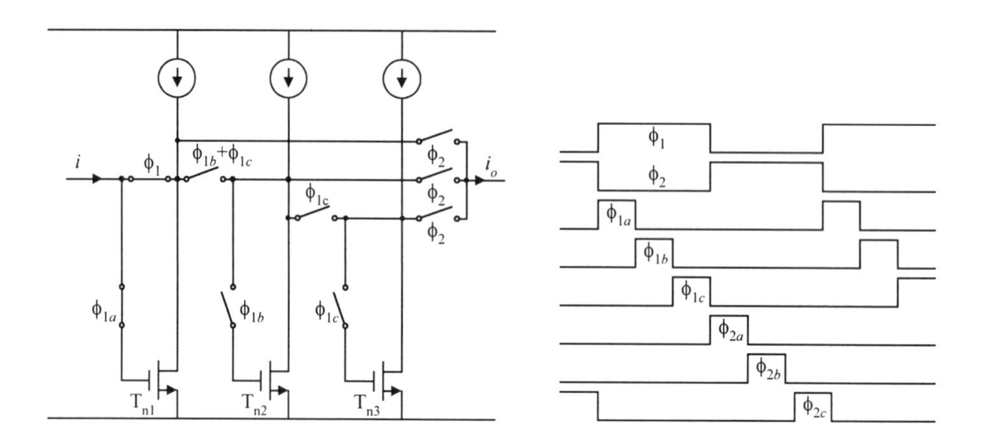

Fig. 3.36. S^nI memory cell with three stages shown.

Therefore, the output current i_o is given by

$$i_o = i_{o1} + i_{o2} + i_{o3} = (-i+\delta_1)+(-\delta_1+\delta_2)+(-\delta_2+\delta_3) = -i+\delta_3 \quad (2.20)$$

where i_{oj} is the output current of the j-th stage, δ_j is the error current generated in the j-th stage, and i is the input current.

Since the input to the first fine memory cell is only the error current generated in the coarse memory cell, and the input to the second fine memory cell is only the error current generated in the first fine memory cell, the total signal transmission error ε is given by

$$\varepsilon = \varepsilon_C \cdot \varepsilon_{F1} \cdot \varepsilon_{F2} \quad (3.21)$$

where ε_C, ε_{F1} , and ε_{F2} are the combined errors of the coarse, the first and the second fine memory cell, respectively. More stages can be cascaded to further reduce the transmission error.

The drawback of the S^nI technique is the relatively complex clocking which may not be desirable for high speed operations.

3.11. SUMMARY

In this chapter, we have detailed different practical SI circuits and circuit techniques including low-distortion SI circuits with the clock feedthrough compensated, low-voltage fully-differential SI circuits, fully-differential class-AB SI circuits, high-speed SI circuits, ultra low-voltage SI circuits, and two-step SI circuits. The CMFF technique and its current-scaled variation have been used in many fully-differential SI circuits in order to eliminate the problems associated with the conventional CMFB technique.

Six test chips were implemented and measured based on the practical SI circuits. For audio applications, the designed first-generation SI memory cell with the clock feedthrough compensated has a THD as low as -65 dBc. The high speed SI circuits can deliver a THD of -47 dBc at a 100-MHz clock frequency with a single 3.3-V supply voltage. For ultra low-voltage applications, the designed SI circuits can deliver a THD of - 48 dBc with a supply voltage of only 1.2 V.

In this chapter, we have also discussed the SI technique in the digital BiCMOS process. By utilizing both bipolar and MOS transistors, we can increase the operation speed and/or increase the accuracy. BiCMOS SI circuits usually have a higher performance and smaller errors than CMOS SI circuits.

Finally, we have touched upon other existing SI techniques which can be utilized to enhance the performance of SI circuits.

Chapter IV: System Design of SI Delta-Sigma Modulators

4.1. Introduction

Due to the development of VLSI technologies, the methodology of system designs has been changed. Traditionally, analog circuits had the lion's share in a signal processing system, especially for filtering applications. The emergence of powerful digital signal processors provided an attractive alternative to these analog implementations, owing to the features of digital signal processors, such as the reduced cost, improved reliability, and enhanced performance, etc. Because the ultimate sources of signals and the final destinations of information are often of the continuous-time (analog) form, the use of powerful digital signal processors to perform the functions that had been done in the analog domain demands high performance in the analog-to-digital (A/D) and digital-to-analog (D/A) converters. At the same time, the voltage range on chips keeps falling to reduce power and increase reliability. And as the circuit densities grow the chip environment becomes more noisy. It is therefore difficult to realize high-performance data converters by conventional methods, especially when the data converters and the digital signal processing (DSP) system are required to be integrated on the same chip, which is advantageous in the design of an application-specific IC (ASIC) chip. The use of oversampling delta-sigma modulating methods thus arises to be a wide acceptance for designing inexpensive but high-performance data converters, oversampling data converters [41].

An oversampling A/D converter usually consists of an analog part and a digital part. The analog part is referred to as the delta-sigma or sigma-delta modulator and the digital part is referred to as the digital decimation filter. However, in the literature, the delta-sigma modulator is sometimes called oversampling A/D converter without reference to the digital decimation filter. It is rather usual to make no distinction between oversampling A/D converters and delta-sigma modulators in the sense that the input to the delta-sigma modulator is analog and the output is digital.

The SC technique has been the dominant circuit technique to realize oversampling delta-sigma modulators [41]. As discussed in the previous chapters, the SI technique is a viable sampled-data technique that can be utilized in oversampling A/D converters. In SI oversampling A/D converters, the advantages of the oversampling modulation technique and the SI technique can be emphasized, while the disadvantages of the SI technique can be eased to a certain extent [16, 17].

Due to the oversampling, the transition band of the anti-aliasing filter is wide and a simple low-order RC filter can meet the requirements for the anti-aliasing. Furthermore, the oversampling reduces the thermal noise power within the signal band, easing the problem of large thermal noise in SI circuits. Oversampling A/D converters using 1-bit quantizers in the noise shaping loop are very robust to inaccuracies in analog components. Linear errors as large as 10% do not degrade the performance significantly [41], overcoming the problem of mismatch in SI circuits.

Oversampling A/D converters usually require very high clock frequencies. The high speed feature of SI circuits attracts. In oversampling A/D converters, high-performance digital circuits are needed. The complete compatibility of SI circuits with the digital CMOS process makes the use of SI circuits an extra bonus. For digital circuits, we need to reduce the supply voltage to save power. SI circuits are suitable for low-voltage operation and therefore, we do not need DC/DC converters to generate a high voltage for a small portion of analog circuits. Applying the SI technique to oversampling A/D converters results in low-cost mixed analog-digital circuits and systems.

In this chapter, we will discuss system design issues, presenting delta-sigma modulator architectures tailored for the SI implementation. Following this short introduction, we will discuss the similarities and differences of the SC and SI implementation. Then we will be concentrating on different SI delta-sigma modulator architectures in this chapter. Also addressed in this chapter is how to increase the dynamic range at a moderate power dissipation at the system level.

4.2. SIMILARITY AND DIFFERENCE OF SC AND SI IMPLEMENTATIONS

Successful designs and implementations require optimization at the system level. In delta-sigma modulators, integrators or differentiators are usually the major building blocks. For the discrete-time implementation, every integrator (or differentiator) should have one sample delay to de-couple the settling chain in order to have a good settling behavior. In this sense, the SC and SI realizations can have the same structure. However, the scaling in SC and SI circuits is different.

In an SC integrator, the scaling factor is directly realized by changing the ratio of the integration capacitor and the sampling capacitor and it has influence on the signal swing within the integrator. Shown in figure 4.1 is a lossless SC integrator.

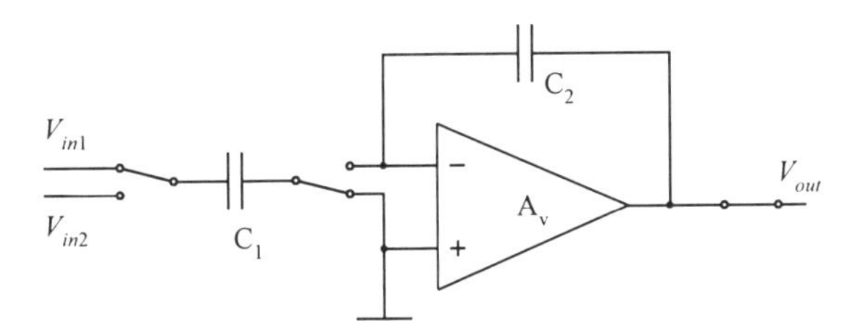

Fig. 4.1. Lossless SC integrator.

The lossless SC integrator shown in figure 4.1 is controlled by a non-overlapping clock. If the output V_{out} is sampled on the same clock phase as the input V_{in1}, we have the following transfer function.

$$V_{out}(z) = \frac{\frac{C_1}{C_2}}{1 - z^{-1}} \cdot \left\{ z^{-1} \cdot V_{in1}(z) - z^{-\frac{1}{2}} \cdot V_{in2}(z) \right\} \tag{4.1}$$

From the input V_{in1} to the output V_{out} there is one clock period delay and from the input V_{in2} to the output V_{out} there is only a half clock period delay. It is of great importance to have delay in the transfer function to de-couple the settling chain.

By changing the capacitance ratio, we can scale the signal. However, things are different in SI integrators. Shown in figure 4.2 is a lossless SI integrator [1] controlled by a non-overlapping clock. The ratios of the transistor sizes and of the bias currents are also shown in the figure.

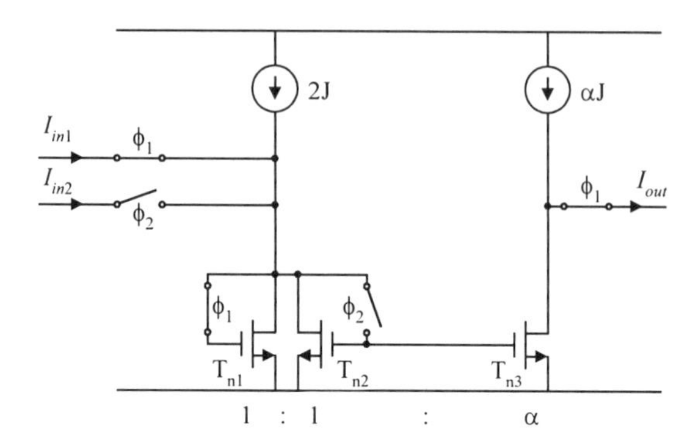

Fig. 4.2. Lossless SI integrator.

If the output I_{out} is only provided on the same clock phase when the input I_{in1} is sampled, we have the following transfer function.

$$I_{out}(z) = \frac{\alpha}{1 - z^{-1}} \cdot \left\{ z^{-1} \cdot I_{in1}(z) - z^{-\frac{1}{2}} \cdot I_{in2}(z) \right\} \tag{4.2}$$

From the input I_{in1} to the output I_{out} there is one clock period delay and from the input I_{in2} to the output I_{out} there is only a half clock period delay. It is of great importance to have delay in the transfer function to de-couple the settling chain.

By changing the size of transistor T_{n3} in respect of the size of transistor T_{n1} (and T_{n2}), different scaling factors can be realized. However, unlike in the SC integrator, the scaling in the SI integrator does not change the signal level within the SI integrator, i.e., the currents in transistors T_{n1} and T_{n2}. Due to this difference, the scaling in SI delta-sigma modulators should differ from their SC counterparts in order to optimize the signal handling capability.

4.3. SECOND-ORDER SI DELTA-SIGMA MODULATORS

The early SI design and implementation did not use a good modulator structure or did not scale properly [42-45]. The modulator structure used in [43] had a poor settling behavior. There were not delays in both integrators, and therefore, the settling of the two integrators and the current quantizer coupled with each other. They had to settle within their final values during the same clock phase and thus the high-frequency performance degraded. A second-order SI delta-sigma modulator by cascading two first-order stages was used in [44], but its serious drawback was the stringent requirement of matching between the digital and SI circuits as pointed out in [41]. The modulators used in [42] and [45] were not optimum since the two integrators had a different range of signal swing.

A general second-order SI delta-sigma modulator structure is shown in figure 4.3. Both integrators have one sample delay. The analog input *x(t)* is a current and the output *y(kT)* is a voltage having CMOS logic levels. The quantizer produces the 1-bit signal by sensing the direction of the current flow from the second integrator. To feedback the quantized signal, *y(kT)* is used as an input to a pair 1-bit D/A converters, one feeding the first integrator and the other feeding the second integrator.

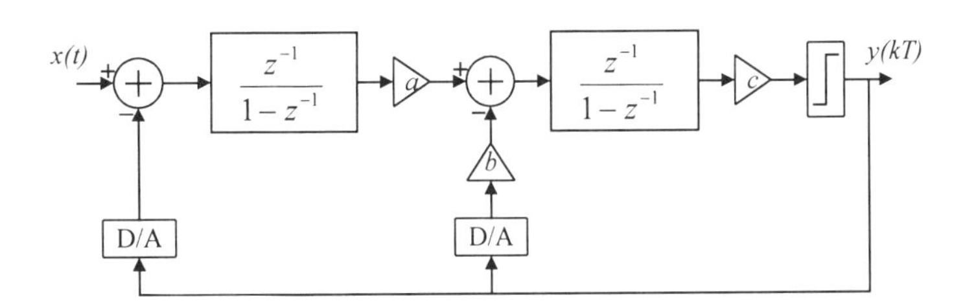

Fig. 4.3. General second-order SI delta-sigma modulator architecture.

In the modulator of figure 4.3, both integrators have a one-sample delay. The settling of the two integrators and the current quantizer is de-coupled. The scaling can be done considering the similarity and difference of SC and SI realizations.

The integrators used in the modulator can be realized by using the configuration shown in figure 4.2. For the negative input (I_{in2} in equation (4.2)), an extra half clock period delay is needed. This extra half clock period delay can be realized in the quantizer which is usually latched.

In an SC integrator, the scaling factor is directly realized by changing the ratio of the integration capacitor and the sampling capacitor. However, in an SI integrator, integration and scaling are done separately as discussed in Chapter 4.2. A scaling factor can either precede or follow the integration. A scaling factor before the first integrator only scales the input signal and the scaling before the second integrator is realized by changing parameters a and b. Due to the stability requirement, b must be equal to 2 times a and their values have an influence on the signal swing at the second integrator output. The factor c does not have any influence on the functionality because the current quantizer only detects the current direction. Modeling the quantizer error as a white noise and considering that the gain of the quantizer is related with the loop gain [51, 52], we have the transfer function given by equation (4.3)

$$Y(z) = z^{-2} \cdot X(z) + \left(1 - z^{-1}\right)^2 \cdot E(z) \tag{4.3}$$

where $Y(z)$, $X(z)$, and $E(z)$ are the digital output, the analog input, and the quantization error, respectively, all expressed in the z domain. Equation (4.3) holds as long as $b = 2a$ and $c > 0$. It is seen from equation (4.3) that a second-order noise shaping loop is realized.

To simulate the delta-sigma modulator of figure 4.3, we use the system-level simulation program TOSCA [46]. The TOSCA software is developed for SC delta-sigma modulators. However, there are other components such as delay elements, adders, multipliers, etc. It is straightforward to construct SI delta-sigma modulators using these components. Nonidealities can also be modeled in these components.

Shown in figure 4.4 is the simulated spectrum. In the simulation, ideal components are assumed. The input signal is a 2-kHz sinusoid with an amplitude 6 dB below the full range (i.e., - 6 dBFS), and the clock frequency is 2.048 MHz. Sixty-four thousand points are used to calculate the FFT. It is seen in figure 4.4 that the slope of the noise floor is 40 dB/dec, confirming the second-order noise shaping. The spectrum output is the same as long as $b = 2a$ and $c > 0$.

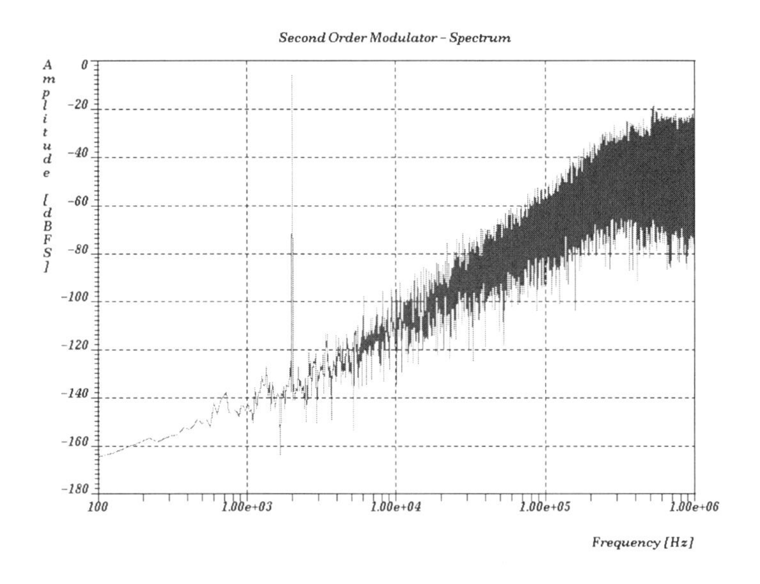

Fig. 4.4. Simulated spectrum of the second-order modulator of Fig. 4.3.

To evaluate the signal-to-noise ratio (SNR), we first need to calculate FFT and find out the signal power and noise power. The noise power can be calculated by integrating all the noise within the signal band. All distortions within the signal band need to be calculated as well. Sometimes it is preferable to calculate all the noise after a digital decimation filter. The general practice is to use a sinc^{L+1} filter (where L is the order of the delta-sigma loop) as the first stage decimation filter [41]. The second stage is usually an FIR filter. For some applications, a single sinc filter suffices as a digital decimation filter for oversampling A/D converters. In the following simulations of the SNRs, we use a sinc^{L+1} filter for an L-th order delta-sigma modulator and the decimation ratio is the same as the oversampling ratio (OSR) if not otherwise stated. The OSR is defined as the sampling frequency vs. the Nyquist frequency.

In figure 4.5, we show the simulated SNR for different input amplitudes. The input signal frequency is 2 kHz and the clock frequency is 2.048 MHz. A $\mathrm{sinc}^3(x)$ filter with a decimation ratio 256 is used to decimate the output of the modulator. The data length for the FFT calculation (in order to evaluate the SNR) is 512. The SNR output is the same as long as $b = 2a$ and $c > 0$.

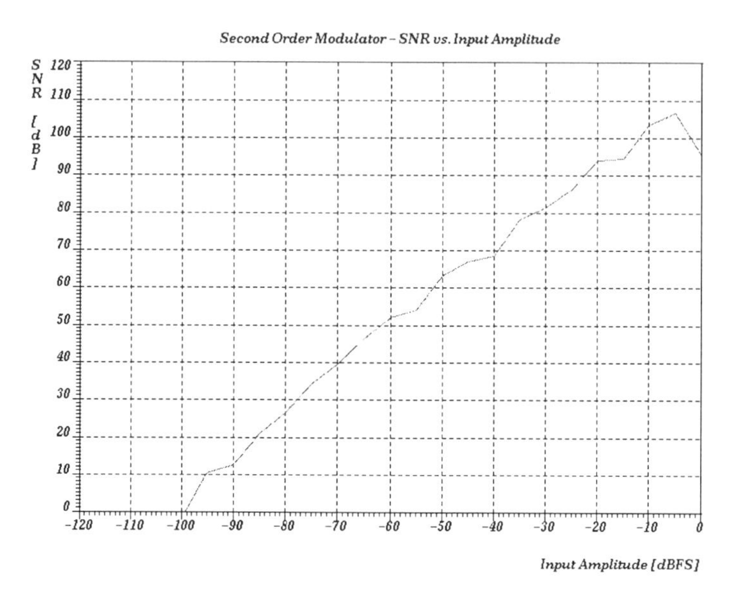

Fig. 4.5. Simulated SNR vs. the input amplitude of the second-order modulator of Fig. 4.3.

Another very important issue is the signal swing within the modulator. In a real implementation, every circuit block has a limited signal processing capability. If the signal swing is larger than what the circuit block can handle, large distortion or even oscillation occurs.

In figure 4.6, we show the simulated histogram of the integrator outputs with $a = 0.5$ and $b = 1$. The input is a 2-kHz sinusoid with an amplitude equal to half of the feedback current value (i.e., - 6 dBFS) and the clock frequency is 2.048 MHz. To calculate the amplitude histogram, 20480 data points are used. We see that both integrators have approximately the same range of signal swing. The modulator (with a = 0.5 and b = 1) only requires a signal range in both integrators slightly larger than twice the full-scale input range (i.e., twice the output current value of the D/A converters). Therefore it is a good candidate for VLSI implementation. Notice that the second integrator has a slightly smaller signal swing than the first one. It is possible to increase the coefficients a and b to increase the signal swing of the second integrator so that both integrators have exactly the same signal swing. However, it might not be rewarding to realize these coefficients due to the increase of irregularity in the layout.

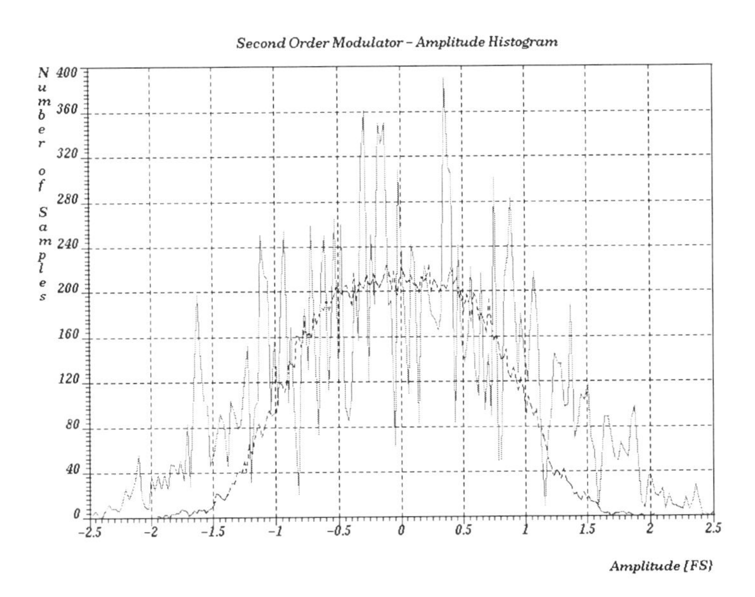

Fig. 4.6. Amplitude histogram of the integrator outputs of the second-order modulator of Fig. 4.3 with a = 0.5 and b = 1. Solid line—the first integrator output; Dotted line—the second integrator output.

If we choose a = 0.25 and b = 0.5 as done in [45], the signal swing of the second integrator output will be as half as that of the first integrator. The signal range of the second integrator will not be utilized. If we choose $a = 1$ and $b = 2$ (this is exactly the case in [42], since the feedback to the second integrator is twice as large as that to the first integrator), the signal swing of the second integrator will be twice as large as that of the first integrator. The second integrator will call for a larger signal range. Therefore, the modulator with $a = 0.5$ and $b = 1$ gives the best performance concerning the signal swing. The choice of the scaling factor c has no effects on the signal swing within the integrators. It only has a slight influence on the design of the current quantizer and on the output current mirror of the second integrator. We can choose, for instance, $c = 1$.

4.4. Two-Stage Fourth-Order SI Delta-Sigma Modulator

For a stable modulator of order L, the SNR increases by $L + 0.5$ bits for each doubling of the oversampling ratio [41]. To achieve a certain SNR, we can either use a high-order modulator structure or use a larger oversampling ratio. For high frequency applications such as in video signal processing, excessively increasing the oversampling ratio is not feasible due to the difficulty in designing very fast VLSI circuits. Therefore, for high-frequency applications, we should use higher-order modulator structures with moderate or even low oversampling ratios.

For higher-order modulators, there exist two basic structures. One is the single-stage, and the other is the multi-stage structure (sometimes is also referred as MASH structure). The single stage structure can usually deliver a very high SNR [47]. However, the input signal magnitude must be well below the feedback signal value not to saturate the modulator. The high SNR comes with the price of large capacitors at the first integrator input in order to reduce the noise floor (if a continuous-time integrator is not used at the input). To stabilize the noise shaping loop, it needs feedforward and/or feedback paths[47]. That means a large capacitance spread and large loads for integrators. All these make the single stage structure not so attractive for high-frequency applications. The multi-stage (usually two-stage) structure cannot deliver a very high SNR due to the matching requirement of analog and digital circuits. However its input signal amplitude is determined by the first stage loop and can be quite large. The multi-stage structure does not have any stability problem and does not need a feedforward or feedback to stabilize it. The structure is also highly regular. Therefore, for high-frequency applications where a very high dynamic range is not required, two-stage modulators are attractive.

In figure 4.7, we show the two-stage fourth-order delta-sigma modulator structure for the SI design and implementation [10]. It consists of two stages, each being a second-order modulator documented for the optimum settling and signal swing [17, 30].

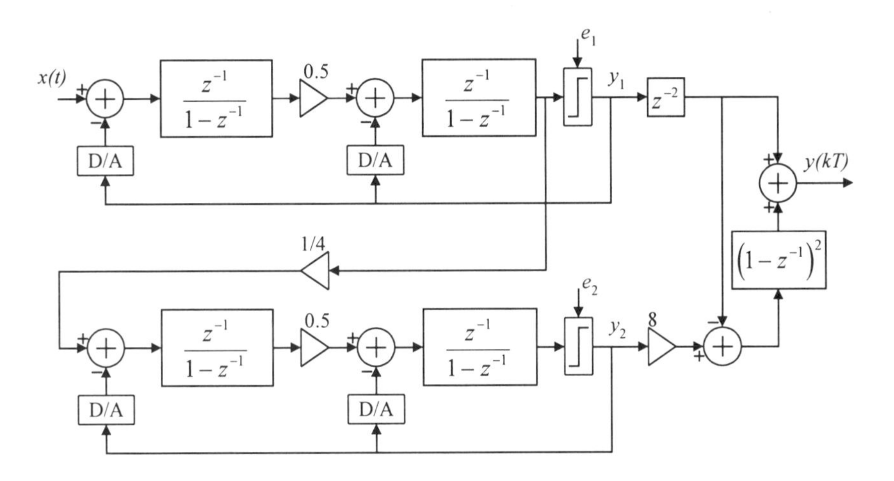

Fig. 4.7. Two-stage fourth-order SI delta-sigma modulator.

The structure is similar to the one proposed for the SC implementation [48, 49]. The difference lies in the scaling due to the consideration of the signal swing [17, 30]. It is also possible to only feed the quantization error of the first stage to the second stage [50], but it requires an instant settling for the quantizer in the first stage and thus is not suitable for high frequency applications. The drawback of the structure of figure 4.7 along with the one proposed in [48, 49] is the larger signal feeding to the second stage compared with the one in [50]. This is overcome by the proper scaling before feeding to the second stage.

Since the output swing of the second integrator in the first stage is approximately two times larger than the full scale signal. In order to keep the same signal swing at the input to the second stage as that at the input to the first stage, a scaling factor of 1/4 is needed as shown in figure 4.7.

Modeling the quantization error as white noise and considering the gain of 2 in the quantizer due to the linearization [51, 52], we have the output $Y_1(z)$ given by

$$Y_1(z) = z^{-2} \cdot X(z) + \left(1 - z^{-1}\right)^2 \cdot E_1(z) \tag{4.4}$$

where $Y_1(z)$ is the digital output of the first stage, $X(z)$ is the analog input, and $E_1(z)$ is the quantization error in the first stage, all expressed in the z domain.

The input to the second stage is the analog output of the first stage. The analog output of the first stage (i.e., the input to the first quantizer) is $\frac{1}{2} \cdot \left\{Y_1(z) - E_1(z)\right\}$. The coefficient 1/2 is due to the gain of the quantizer (equal to 2 in this particular case) when we use the linearized model [51, 52]. The input to the second stage is $\frac{1}{8} \cdot \left\{Y_1(z) - E_1(z)\right\}$. Modeling the quantization error as white noise and considering the gain of 2 in the second quantizer due to the linearization, we have the output of the second stage $Y_2(z)$ given by

$$\begin{aligned} Y_2(z) &= z^{-2} \cdot \frac{1}{8} \cdot \left\{Y_1(z) - E_1(z)\right\} + \left(1 - z^{-1}\right)^2 \cdot E_2(z) \\ &= z^{-4} \cdot \frac{1}{8} \cdot X(z) + z^{-2} \cdot \frac{1}{8} \cdot \left\{\left(1 - z^{-1}\right)^2 - 1\right\} \cdot E_1(z) \\ &+ \left(1 - z^{-1}\right)^2 \cdot E_2(z) \end{aligned} \tag{4.5}$$

where $Y_2(z)$ is the digital output of the second stage, $Y_1(z)$ is the digital output of the first stage, $E_1(z)$ is the quantization error in the first stage, $E_2(z)$ is the quantization error in the second stage, and $X(z)$ is the analog input, all expressed in the z domain.

To cancel the quantization error introduced in the first stage $E_1(z)$, some processing is needed in the digital domain as shown in figure 4.7. The final output $Y(z)$ is given by

$$\begin{aligned} Y(z) &= z^{-2} \cdot Y_1(z) + \left(1 - z^{-1}\right)^2 \cdot \left\{8Y_2(z) - z^{-2} \cdot Y_1(z)\right\} \\ &= z^{-4} \cdot X(z) + 8\left(1 - z^{-1}\right)^4 \cdot E_2(z) \end{aligned} \tag{4.6}$$

It is seen that the quantization error of the first stage is canceled by the following stage and that the structure of figure 4.7 realizes a fourth-order noise shaping function.

In figure 4.8, we show a typical output spectrum with ideal components. The input signal frequency is 2 kHz, the input signal amplitude is 6 dB below the full scale (i.e., - 6 dBFS), and the clock frequency is 2.048 MHz. We use 64-k points to calculated the FFT and a Harris window is also used. It is seen from figure 4.8 that the noise floor increases at 80 dB/dec. The architecture realizes a fourth-order noise shaping.

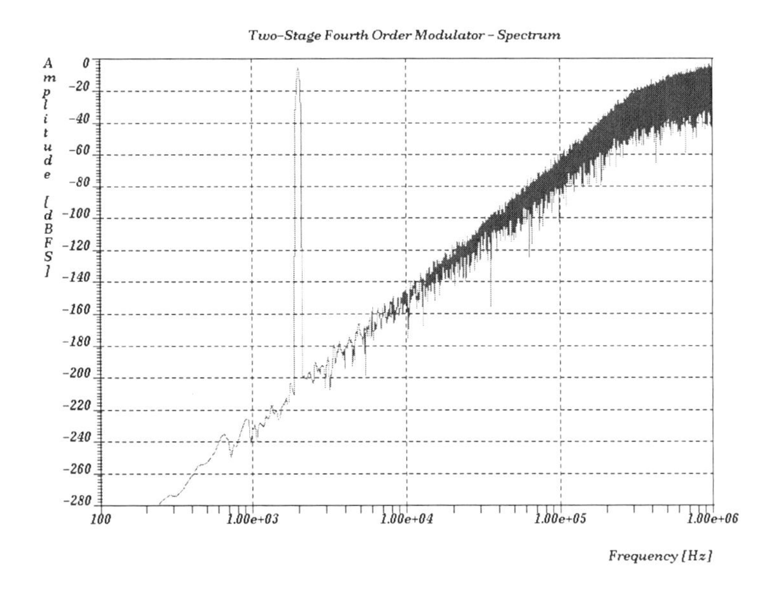

Fig. 4.8. Simulated spectrum of the two-stage fourth-order modulator of Fig. 4.7.

However, the mismatch between the analog and digital circuits limits the performance. Since the mismatch error only results in an error that has already been subject to a second-order noise shaping, the mismatch is not devastating to the performance. We show the effect in figure 4.9. The input signal is a 400-kHz sinusoid and the clock frequency is 40.96 MHz. The decimator filter is $\text{sinc}^5(x)$ with a decimation ratio of 16 and 512 points are used to calculate the FFT.

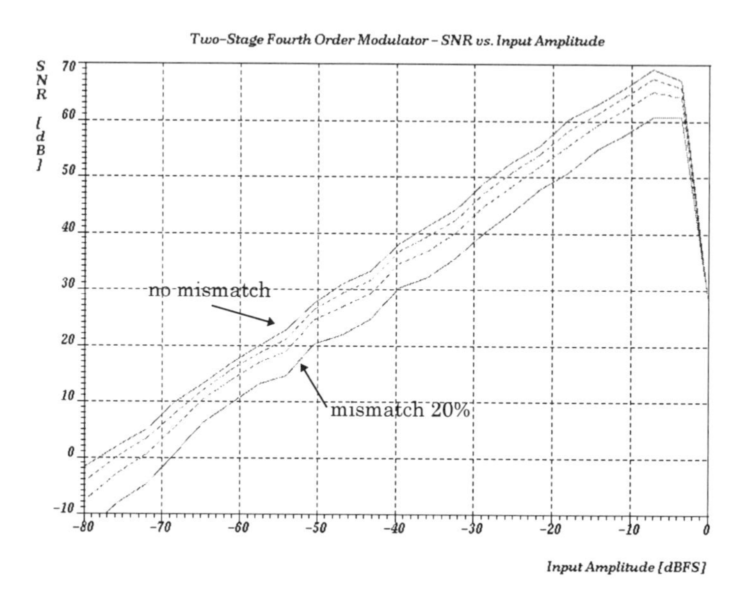

Fig. 4.9. Simulated SNR vs. input signal amplitude for different mismatch errors. The mismatch error (from upper to lower) is 0%, 5%, 10%, and 20%, respectively.

It is seen that without the mismatch error, the fourth-order modulator of figure 4.7 can deliver a dynamic range of 78 dB with an oversampling ratio of 16 (notice that the peak SNR is ~70 dB). When a mismatch error between the analog and digital circuits is introduced, the performance degrades due to the incomplete canceling of the quantization error of the first stage. With mismatch error as large as 20%, however, the modulator can still achieve a dynamic range of about 68 dB with an oversampling ratio of 16. Therefore, the modulator of figure 4.7 is well suited for high-frequency applications.

We show in figure 4.10 the amplitude histogram of integrator outputs of the second stage of the two-stage fourth-order delta-sigma modulator of figure 4.7. (The histogram of the first stage is identical to that of the modulator of figure 4.3.) The input signal is a 2-kHz sinusoid with an amplitude equal to half of the feedback signal value (i.e., - 6 dBFS) and the clock frequency is 2.048 MHz. To plot the histogram, 20480 data are used.

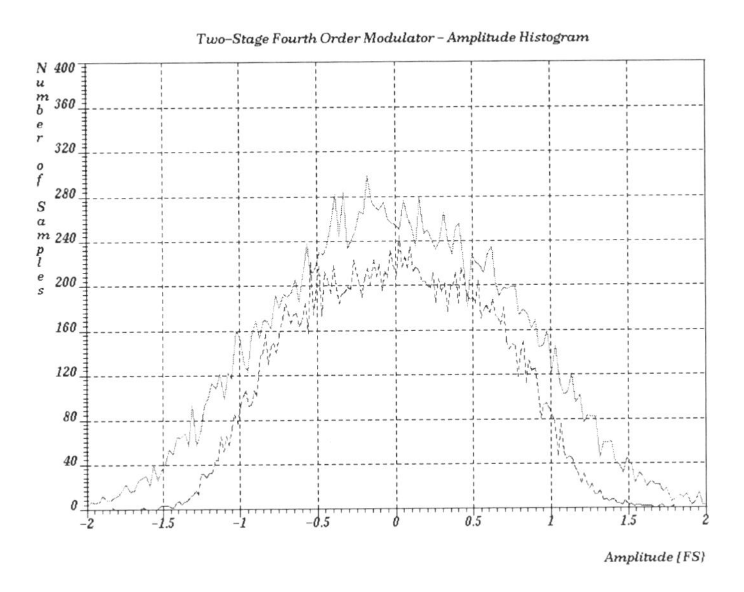

Fig. 4.10. Histogram of integrator outputs of the second stage of the two-stage fourth-order SI modulator of Fig. 4.7. Solid line—the first integrator output; Dotted line—the second integrator output.

It is seen that the two-stage fourth-order SI delta-sigma modulator of figure 4.7 only requires a signal range within all the integrators slightly larger than twice the feedback signal value. Therefore, it is a good candidate for the VLSI implementation. Notice that the second integrator has a slightly smaller signal swing than the first one. It is possible to increase the signal swing of the second integrator so that both integrators have exactly the same signal swing. However, it might not be rewarding to realize these coefficients due to the increase of irregularity in the layout.

4.5. Single-Stage Fourth-Order Delta-Sigma Modulator

It is also possible to realize higher-order delta-sigma modulators only by using one stage. Single-stage higher-order delta-sigma modulators are better candidates for higher resolution applications than multi-stage higher-order delta-sigma modulators [47], though for mediate-resolution high-speed applications multi-stage structures are a better choice as discussed in Chapter 4.4.

To stabilize a single-stage higher-order (>2) delta-sigma modulator, the poles must be moved inside the unit circle [41]. Sometimes zeroes need to be introduced into the noise shaping function. Therefore, the attenuation of the quantization noise at low frequencies in a single-stage higher-order (>2) modulator is less than that in a multi-stage modulator having the same order of the noise shaping function. The slope of the noise floor at high frequencies of a single-stage higher-order modulator is also less than that of a multi-stage modulator having the same order of the noise shaping function.

In figure 4.11 we show a single-stage fourth-order SI delta-sigma modulator, it is based on an SC realization [53].

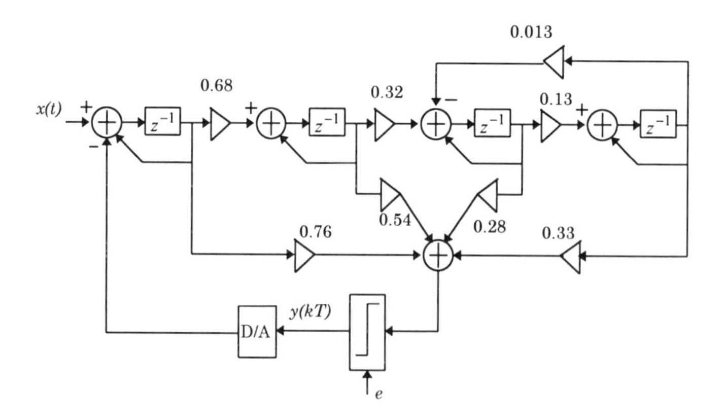

Fig. 4.11. Single-stage fourth-order SI delta-sigma modulator.

Four integrators are used in this modulator with multiple feedforward paths to the quantizer and a global feedback to introduce a zero in the noise shaping function. As seen in figure 4.11, there is a wide spread of coefficients. It usually has a negative effect on the speed in a realization. In the realization, unit transistors are usually used. Some coefficients need many unit transistors in parallel due to the wide spread in coefficients. This decreases the speed or increases the power. (Speed may be maintained by increasing the transconductance and therefore the power increases.) Very small coefficients can be realized by several stages by cascading current mirrors. Due to the settling chain, it usually degrades the speed as well. Another drawback is the irregularity in the layout due to the wide spread of coefficients.

In figure 4.12, we show the simulated spectrum. The input signal frequency is 2 kHz, the input signal amplitude is 6 dB below the full scale (i.e., - 6 dBFS), and the clock frequency is 2.048 MHz. We use 64 K points to calculated the FFT and a Harris window is also used.

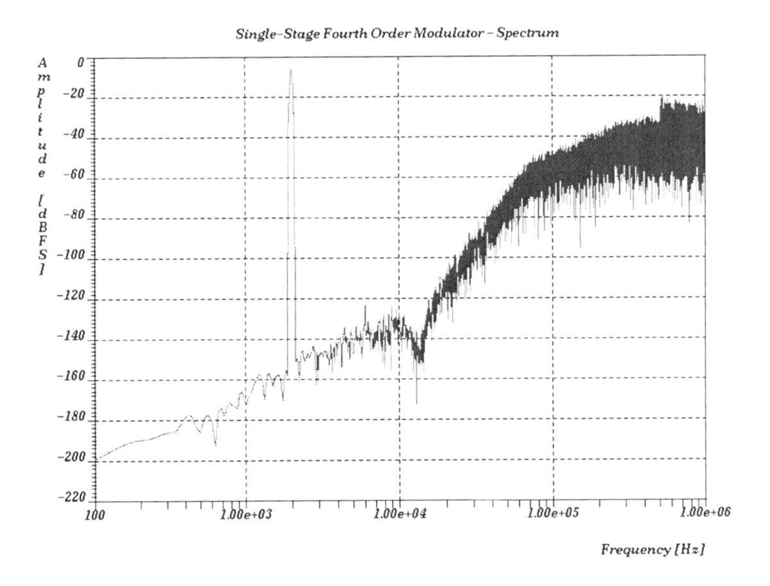

Fig. 4.12. Simulated spectrum of the single-stage fourth-order modulator of Fig. 4.11.

It is evident in figure 4.12 that there is a zero in the noise transfer function and that the attenuation of the quantization noise at low frequencies is less than that in the two-stage fourth-order modulator of figure 4.7. The slope of the noise floor at high frequencies is also less than that of the two-stage fourth-order modulator of figure 4.7 (see figure 4.8).

In figure 4.13, we show the simulated SNR for different input amplitudes. The input signal is a 400-kHz sinusoid and the clock frequency is 40.96 MHz. The decimator filter is $\mathrm{sinc}^5(x)$ with a decimation ratio of 16 and 512 points are used to calculate the FFT. The solid line is the simulated SNR without coefficient errors and the dotted line is the simulated SNR with 10% coefficient errors. It is seen in the figure that the single-stage fourth-order delta-sigma modulator architecture of figure 4.11 is not sensitive to coefficient errors.

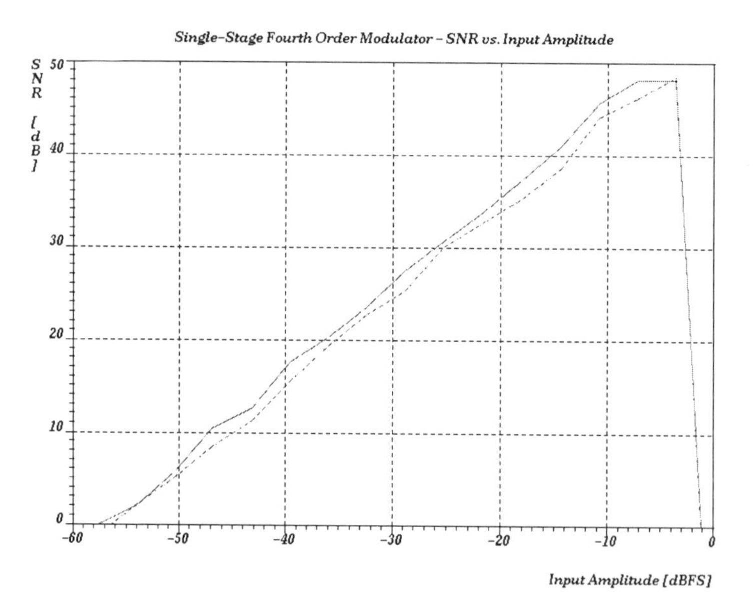

Fig. 4.13. Simulated SNR vs. input amplitude of the single-stage fourth-order SI delta-sigma modulator of Fig. 4.11. Solid line: no coefficient errors; Dotted line: 10% coefficient errors.

Compared with the simulation result of the two-stage fourth-order delta-sigma modulator of figure 4.7, the single-stage fourth-order delta-sigma modulator of figure 4.11 delivers a lower dynamic range due to the moving of poles inside the unit circle and the introduction of the zero in the noise shaping function. However, the single-stage fourth-order delta-sigma modulator of figure 4.11 does not call for matching of digital and analog parts. It is also seen by comparing figures 4.9 and 4.13 that the SNR of the single-stage modulator of figure 4.11 drops more rapidly when the input signal amplitude increases due to the clipping of signals. Due to all these reasons, the two-stage fourth-order delta-sigma modulator of figure 4.7 is usually preferred over the single-stage fourth-order delta-sigma modulator of figure 4.11 for SI implementation, especially for high speed applications.

In figure 4.14, we show the simulated histogram of the integrator outputs of the single-stage fourth-order delta-sigma SI modulator of figure 4.11. The input signal is a 2-kHz sinusoid with an amplitude equal to half of the feedback current value (i.e., - 6 dBFS) and the clock frequency is 2.048 MHz. To calculated the amplitude histogram, 20480 data points are used.

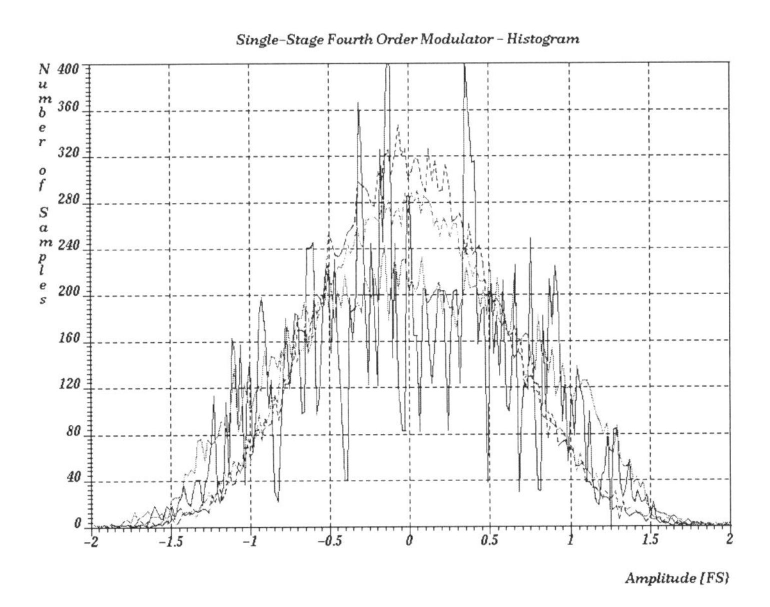

Fig. 4.14. Simulated histogram of the single-stage fourth-order modulator of Fig. 4.11. All the integrators have the same signal swing.

It is evident from figure 4.14 that all the four integrators have the same range of signal swing. All the integrators need to handle a signal range of twice the feedback current. Therefore, the single-stage fourth-order delta-sigma modulator of figure 4.11 is a good candidate for the SI implementation.

4.6. CHOPPER-STABILIZED SI DELTA-SIGMA MODULATOR

For audio applications, the low-frequency noise such as flicker noise may limit the dynamic ranges. We can reduce the low-frequency noise by the correlated double sampling or chopper stabilization techniques. (Notice that the correlated double sampling is inherent in the second-generation SI circuits, see Chapter 2.7.) Although the correlated double sampling or chopper stabilization techniques can be applied to a single amplifying stage to reduce the low-frequency noise, the chopper stabilization at the system level can reduce the circuit noise more efficiently [54]. A chopper stabilized modulator can be derived from the existing modulator by applying the transformation $z \longrightarrow -z$ [54]. Considering the difference and similarity of SC and SI realizations, we show a chopper stabilized second-order SI delta-sigma modulator [33, 34] in figure 4.15. It is derived from the modulator of figure 4.3.

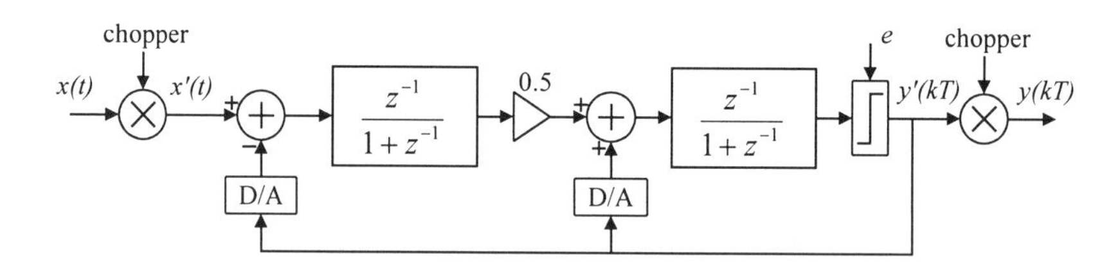

Fig. 4.15. Chopper-stabilized SI delta-sigma modulator.

The input *x(t)* is first chopper stabilized before fed to the core modulator. Its spectrum is moved to high frequencies around half of the sampling frequency. Therefore, the low-frequency noise at the core modulator input has no way to contaminate the input signal. The core modulator is designed to have a high-pass signal transfer function and a low-pass noise shaping function. At the modulator output, a chopper multiplication is used to move the signal spectrum to low frequencies and thus generate the desired digital output. If low-frequency noise is introduced at the input of the core modulator, it will be attenuated by the high-pass signal transfer function, having little effect on the input signal that has been moved to high frequencies due to the input chopper stabilization.

Now, let's derive the transfer function. When we chopper stabilize a signal, we change the sign of the signal alternately and maintain the amplitude. At even (or odd) samples we do not change the sign of the input signal while at the odd (or even) samples we change the sign of the input signal. This is equivalent to moving the spectrum in respect to the sampling frequency, therefore we have

$$X(z) = X'(-z) \tag{4.7}$$

and

$$Y(z) = Y'(-z) \tag{4.8}$$

where $X(z)$ is the analog input signal, $X'(z)$ is the analog input signal after the chopper stabilization, $Y(z)$ is the digital output signal, and $Y'(z)$ is the modulator digital output before the chopper stabilization, all expressed in the z domain.

Considering the gain factor 2 in the quantizer when using the linearized model, we have the following equation

$$Y'(z) = z^{-2} \cdot X'(z) + \left(1 + z^{-1}\right)^2 \cdot E(z) \tag{4.9}$$

where $Y'(z)$ is the modulator output before the chopper stabilization, $X'(z)$ is the analog input after the chopper stabilization, and $E(z)$ is the quantization error introduced in the quantizer, all expressed in z domain.

Combining equations (4.7) ~ (4.9), we have

$$\begin{aligned} Y(z) &= Y'(-z) = (-z)^{-2} \cdot X'(-z) + \left(1 + (-z)^{-1}\right)^2 \cdot E(-z) \\ &= z^{-2} \cdot X(z) + \left(1 - z^{-1}\right)^2 \cdot E(-z) \end{aligned} \tag{4.10}$$

Since the quantization error can be modeled as white noise, therefore we have

$$E(z) = E(-z) \tag{4.11}$$

Then equation (4.10) becomes

$$Y(z) = z^{-2} \cdot X(z) + \left(1 - z^{-1}\right)^2 \cdot E(z) \tag{4.12}$$

It is seen from equations (4.12) and (4.3) that the chopper-stabilized delta-sigma modulator of figure 4.15 realizes the same signal transfer and noise shaping function as the delta-sigma modulator of figure 4.3.

In figures 4.16 and 4.17, we show the simulated output spectra of the chopper-stabilized modulator outputs calculated by a 16-k-point FFT. The input signal is a 2-kHz sinusoid with a - 6 dBFS amplitude and the clock frequency is 2.048 MHz. To evaluate the effectiveness of the chopper stabilization, we also add another 1-kHz sinusoid with a - 20 dBFS amplitude at the input of the core modulator of figure 4.15 (after the input chopper stabilization), emulating the low-frequency disturbance.

Shown in figure 4.16 is the simulated modulator output spectrum before the output chopper multiplication. The signal spectrum is moved to the band close to half of the sampling frequency due to the input chopper stabilization. In this case, it is at 1.024 MHz - 2 kHz = 1.022 MHz. It is not suppressed by the high-pass signal transfer function of the core delta-sigma modulator. The 1-kHz disturbance introduced at the input of the core modulator after the input chopper stabilization is not moved to high frequencies and it is consequently suppressed by the high-pass signal transfer function of the core delta-sigma modulator.

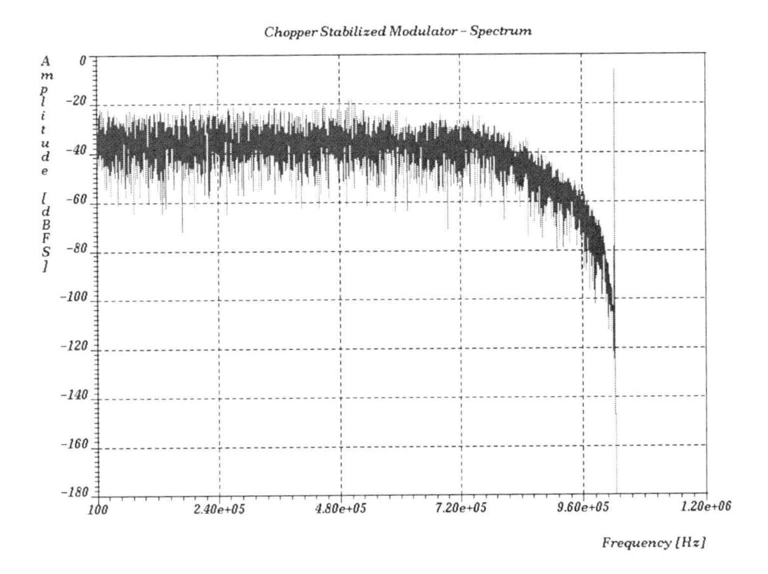

Fig. 4.16. Simulated spectrum of the output before the output chopper stabilization of the chopper-stabilized SI delta-sigma modulator of Fig. 4.15.

Show in figure 4.17 are the output spectrum of the chopper-stabilized SI delta-sigma modulator of figure 4.15 (solid line) and the output spectrum of the ordinary SI delta-sigma modulator of figure 4.3 (dotted line). A 1-kHz -20-dBFS signal is also added at the input of the non-chopper-stabilized delta-sigma modulator of figure 4.3. It is seen in figure 4.17 that the 1-kHz disturbance is attenuated thanks to the chopper stabilization in the chopper-stabilized SI delta-sigma modulator of figure 4.15. Without the chopper stabilization, the 1-kHz disturbance is clearly in the signal band as shown by the dotted line in figure 4.17. Therefore, the chopper-stabilized SI delta-sigma modulator is not sensitive to low frequency noise introduced at the input of the core modulator.

It is also evident from figure 4.17 that the SI delta-sigma modulators of figures 4.15 and 4.3 have the same noise shaping function.

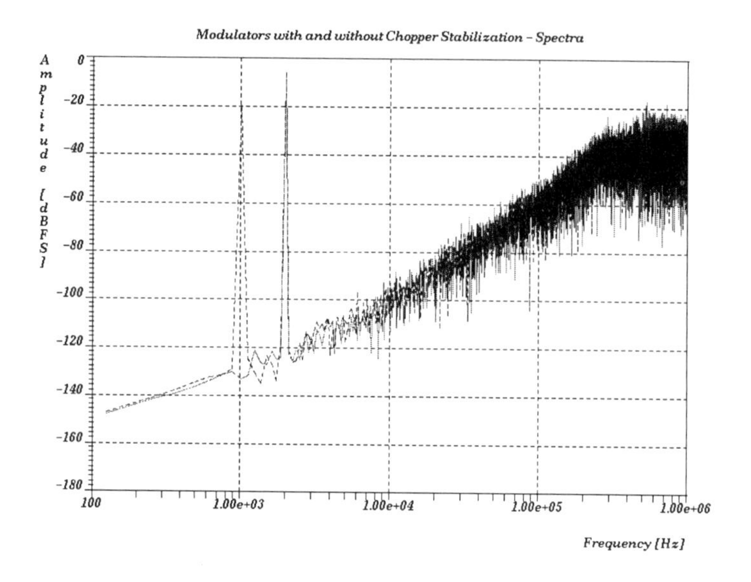

Fig. 4.17. Simulated spectra of the SI delta-sigma modulators with and without chopper stabilization. Solid line: output of the chopper-stabilized modulator of Fig. 4.15; Dotted line: output of the ordinary delta-sigma modulator of Fig. 4.3.

In figure 4.18, we show the simulated histogram. The input signal is a 2-kHz sinusoid with a -6 dBFS amplitude and the clock frequency is 2.048 MHz. We use 20480 data points to plot the histogram. We see that both differentiators have the approximately same range of signal swing. The chopper-stabilized modulator only requires a signal range in both differentiators slightly larger than twice the full-scale input range. Therefore, the chopper-stabilized modulator of figure 4.15 is a good candidate for the VLSI implementation where the signal range is restricted. Notice that the second differentiator has a slightly smaller signal swing than the first one. It is possible to increase the signal swing of the second differentiator so that both differentiators have exactly the same signal swing. However, it might not be rewarding to realize these coefficients due to the increase of irregularity in the layout.

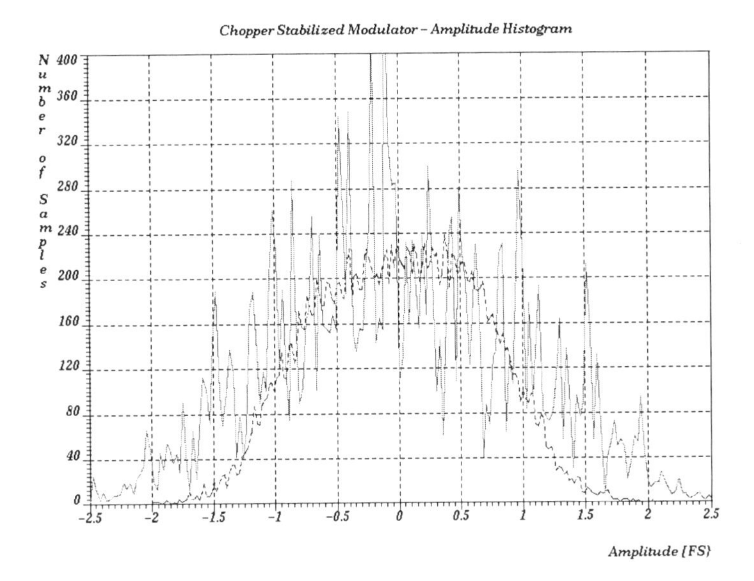

Fig. 4.18. Histogram of the differentiator outputs of the chopper-stabilized SI delta-sigma modulator of Fig. 4.15.
Solid line—the first differentiator output;
Dotted line—the second differentiator output.

4.7 Current Scaling Technique for SI Delta-Sigma Modulators

A. Current scaling principle for SI delta-sigma modulators

Low frequency noise such as flicker noise is reduced in the second generation SI circuits due to the correlated double sampling as discussed in Chapter 2.7. Thermal noise at the input usually imposes a fundamental limitation on the dynamic range of SI delta-sigma modulators. To increase the dynamic range, we have to increase the power dissipation. As discussed in Chapter I, for every doubling of the bias current, we can increase the dynamic range by 3 dB for a thermal-noise limited SI circuit. By applying this principle, we can conclude that we can increase the dynamic range by 3 dB for a thermal-noise limited SI delta-sigma modulator by doubling the power dissipation. Obvious this is costly.

In an oversampling delta-sigma modulator, however, only the thermal noise at the first integrator input limits the achievable dynamic range. The noise introduced at the following integrators experiences a noise shaping due to the modulation loop and therefore it usually does not limit the dynamic range. This observation leads to the current scaling principle for oversampling SI delta-sigma modulators.

To achieve a high dynamic range in a thermal-noise limited SI delta-sigma modulator, we can design the first integrator (or differentiator) with a very large bias current to increase the full-scale input current. The output of the first integrator (or differentiator) is then scaled by a large factor. Since the input currents to the following stages are small, the integrators (or differentiators) in the following stages can be biased at relatively low currents to reduce the power dissipation and chip area.

To maintain the signal graph property (transfer function), all the other coefficients need to be scaled proportionally as in a linear system. The only difference between the scaling in a linear system and the scaling in a delta-sigma modulator having 1-bit current quantizers is that the scaling factors immediately preceding the 1-bit current quantizers do not change the property of the system. The reason is that the 1-bit current quantizers only detect the current flow direction.

Also noted is that the use of different bias currents in different integrators (or differentiators) does not have any influence on the speed if both bias currents and transistor width are scaled as discussed in Chapter I. Therefore, the current scaling principle does not degrade the operation speed of SI delta-sigma modulators.

This is best illustrated by the following examples [55].

B. Current-scaled two-stage fourth-order SI delta-sigma modulator

Shown in figure 4.19 is a two-stage fourth-order SI delta-sigma modulator with the current-scaled. It is based on the architecture of figure 4.7.

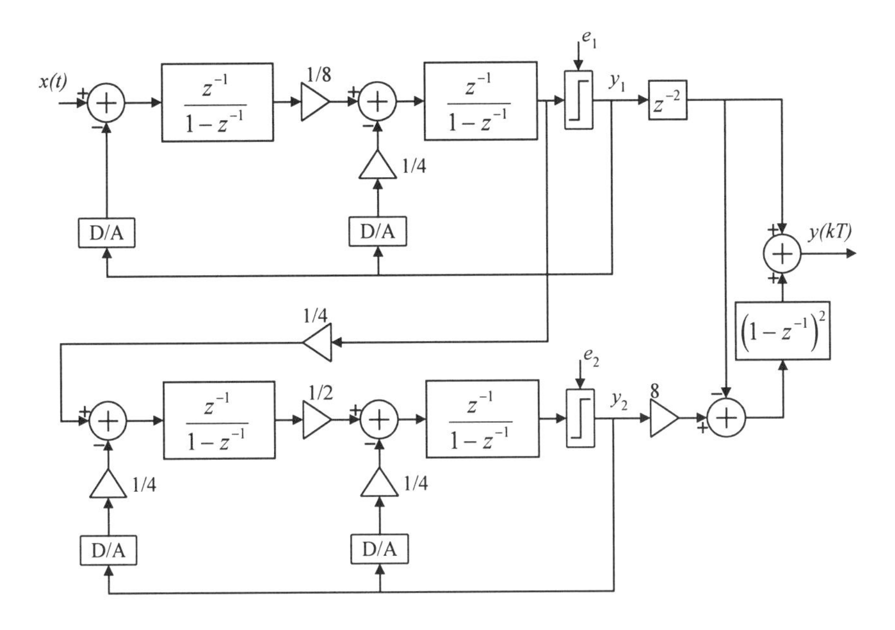

Fig. 4.19. Two-stage fourth-order SI delta-sigma modulator with the current scaled.

The output of the first integrator in the first stage is further scaled by a factor of 4 compared with the SI delta-sigma modulator of figure 4.7. Therefore, the feedback to the second integrator needs to be scaled by a factor of 4. To guarantee the signal flow graph property, the output of the second integrator has to be increased by a factor of 4. Since this factor precedes the 1-bit current quantizer, we can change it to unity without changing the signal transfer function. The same principle applies to the second stage as well. Thereby the current-scaled SI delta-sigma modulator shown in figure 4.19 results. Since the signal is scaled down after the first integrator and the input current to the second integrator of the first stage is reduced, a smaller bias current can be used for the second integrator to save power. For the second stage, small bias currents can also be used to save power and chip area since the input current to the second stage is scaled down.

Since we apply the scaling principle of a linear system to the delta-sigma modulator of figure 4.7, the transfer function should be same as given by equation (4.6). We can also derive the transfer function directly from figure 4.19.

Modeling the quantization error as white noise and considering the quantizer has a gain of 8 with the linearization [51, 52], we have the output $Y_1(z)$ given by

$$Y_1(z) = z^{-2} \cdot X(z) + \left(1 - z^{-1}\right)^2 \cdot E_1(z) \tag{4.13}$$

where $Y_1(z)$ is the digital output of the first stage, $X(z)$ is the analog input, and $E_1(z)$ is the quantization error in the first stage, all expressed in the z domain.

The input to the second stage is the analog output of the first stage. The analog output of the first stage (i.e., the input to the first quantizer) is $\frac{1}{8} \cdot \{Y_1(z) - E_1(z)\}$. The coefficient 1/8 is due to the gain of the quantizer (equal to 8 in this particular case) when we use the linearized model [51, 52]. Therefore, the input to the second stage is $\frac{1}{32} \cdot \{Y_1(z) - E_1(z)\}$.

Since the feedback to the first integrator in the second stage is scaled by a factor of 4, it has the same effect as the input is increased by a factor of 4 when we apply the linearized model [51, 52]. Therefore, the output of the second stage $Y_2(z)$ is given by

$$\begin{aligned} Y_2(z) &= z^{-2} \cdot 4 \cdot \left\{ \frac{1}{32} \cdot \{Y_1(z) - E_1(z)\} \right\} + \left(1 - z^{-1}\right)^2 \cdot E_2(z) \\ &= z^{-4} \cdot \frac{1}{8} \cdot X(z) + z^{-2} \cdot \frac{1}{8} \cdot \left\{ \left(1 - z^{-1}\right)^2 - 1 \right\} \cdot E_1(z) \\ &\quad + \left(1 - z^{-1}\right)^2 \cdot E_2(z) \end{aligned} \tag{4.14}$$

where $Y_2(z)$ is the digital output of the second stage, $Y_1(z)$ is the digital output of the first stage, $E_1(z)$ is the quantization error in the first stage, $E_2(z)$ is the quantization error in the second stage, and $X(z)$ is the analog input, all expressed in the z domain.

To cancel the quantization error $E_1(z)$ introduced in the first stage, some processing is needed in the digital domain as shown in figure 4.19. The final output $Y(z)$ is given by

$$\begin{aligned} Y(z) &= z^{-2} \cdot Y_1(z) + \left(1 - z^{-1}\right)^2 \cdot \left\{8Y_2(z) - z^{-2} \cdot Y_1(z)\right\} \\ &= z^{-4} \cdot X(z) + 8\left(1 - z^{-1}\right)^4 \cdot E_2(z) \end{aligned} \tag{4.15}$$

It is seen that the transfer function of the current-scaled SI delta-sigma modulator of figure 4.19 is exactly the same as the one of figure 4.7. The scaling does not have any influence on the signal transfer and noise shaping functions and the structure of figure 4.19 realizes a fourth-order noise shaping function. This can also be seen in the simulated spectrum shown in figure 4.20.

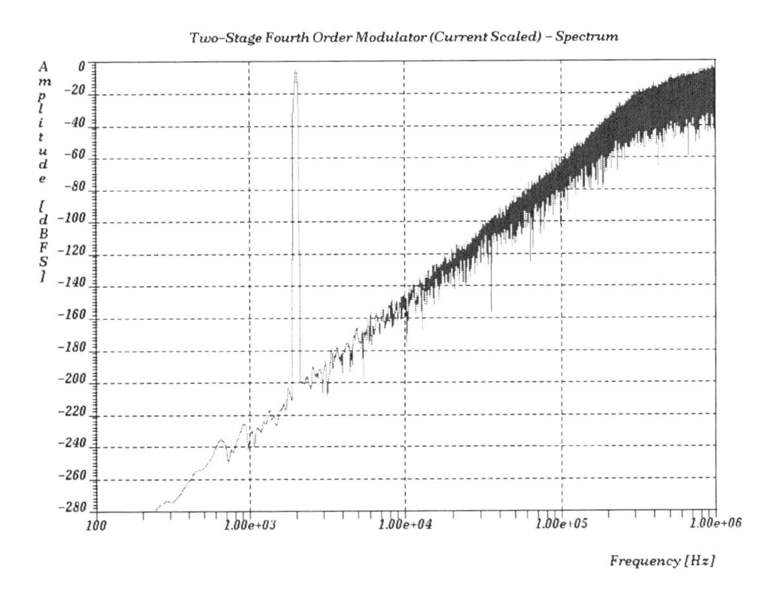

Fig. 4.20. Simulated output spectrum of the current-scaled two-stage fourth-order delta-sigma modulator of Fig. 4.19.

Shown in figure 4.20 is the simulated spectrum of the current-scaled two-stage fourth-order delta-sigma modulator of figure 4.19. It is exactly the same as the one shown in figure 4.8. The input signal frequency is 2 kHz, the input signal amplitude is 6 dB below the feedback current to the first integrator (i.e., -6 dBFS), and the clock frequency is 2.048 MHz. The spectrum is calculated by a 64-k-point FFT preceded by a Harris window. It is seen in figure 4.20 that the slope of the noise floor is 80 dB/dec. The architecture realizes a fourth-order noise shaping.

Just as the delta-sigma modulator of figure 4.7, the current-scaled delta-sigma modulator of figure 4.19 is also sensitive to the mismatch between the analog and digital circuits in that the cancellation of the quantization error of the first stage relies on the matching of the analog and digital circuits. Since the mismatch error only results in an error that has already been subject to a second-order noise shaping, the mismatch is not devastating to the performance. The mismatch effect on the performance of the current-scaled modulator of figure 4.19 is the same as for the modulator of figure 4.7 as illustrated in figure 4.9. Both modulators have the identical behavior.

However, the signal swing is different in the current-scaled delta-sigma modulator of figure 4.19 compared with the one of figure 4.7. Unlike in the delta-sigma modulator of figure 4.7 where every integrator has the same signal swing, we scale down the signal swings after the first integrator in the new architecture. In figure 4.21, we show the simulated amplitude histogram. The input signal frequency is 2 kHz, the input signal amplitude is 6 dB below the feedback current to the first integrator (i.e., 6 dBFS), and the clock frequency is 2.048 MHz. We use 20480 data to plot the histogram.

It is seen from figure 4.21 that the signal swing within the first integrator is almost four times as large as the signal swings within all the other integrators. Therefore, the bias current in the first integrator can be designed 4 times as large as the rest integrators. The dynamic range can be increased without a drastic increase in the power dissipation or chip area. If we use the architecture of figure 4.7 to achieve the same dynamic range, the power dissipation and chip area will be more than two times larger than the one shown in figure 4.19. (Assume that the thermal noise is the major limitation of the dynamic range.)

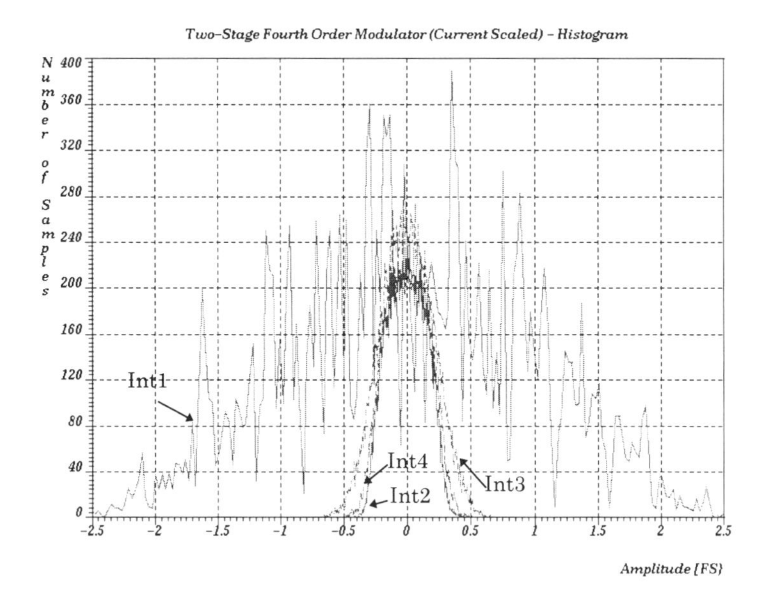

Fig. 4.21. Histogram of the integrator outputs of the current-scaled two-stage fourth-order delta-sigma modulator of figure 4.19.

C. Current-scaled single-stage fourth-order SI delta-sigma modulator

Another example is a single-stage fourth-order modulator as shown in figure 4.22. It is based on the SI delta-sigma modulator shown in figure 4.11.

We first scale down the first output of the first integrator (i.e., the input to the second integrator) and maintain the second output of the first integrator (i.e., one of the quantizer's inputs). To maintain the signal flow graph property, both outputs of the second integrator have to be scaled up by the same factor. Then we re-scale the output of the second integrator and so on. The four inputs to the quantizer (the adder immediately preceding the quantizer) can be scaled arbitrarily by the same factor without changing the system behavior.

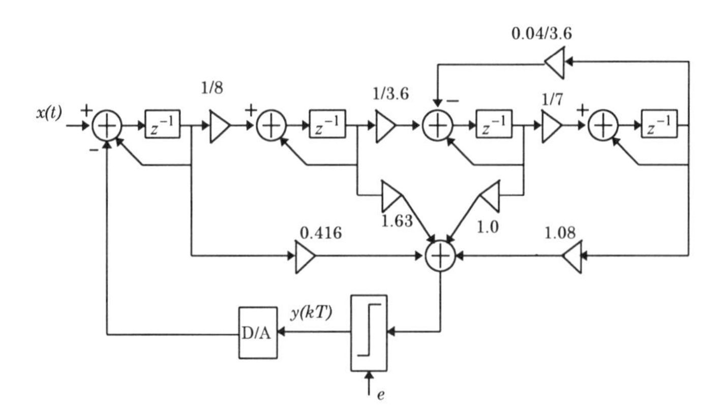

Fig. 4.22. Single-stage fourth-order delta-sigma modulator with the current scaled.

Shown in figure 4.23 is the simulated spectrum. The input signal frequency is 2 kHz, the input signal amplitude is 6 dB below the feedback current to the first integrator (i.e., -6 dBFS), and the clock frequency is 2.048 MHz. The spectrum is calculated by a 64-k-point FFT preceded by a Harris window.

It is seen from figure 4.23 that the modulator of figure 4.22 realizes a single-stage fourth-order noise shaping function with a transmission zero. It is almost identical to the one shown in figure 4.12. The minor difference is due to the coefficient truncation when we scale the delta-sigma modulator of figure 4.11.

Just as the delta-sigma modulator of figure 4.11, the current-scaled delta-sigma modulator of figure 4.22 is also not so sensitive to the coefficient errors. The effect of coefficient errors on the performance of the current-scaled modulator of figure 4.22 is the same as for the modulator of figure 4.11 as illustrated in figure 4.13. Both modulators have a similar behavior.

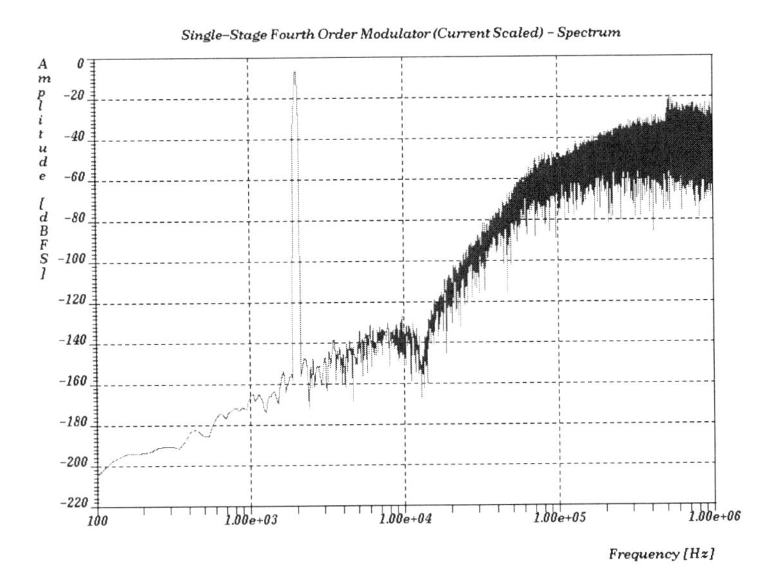

Fig. 4.23. Simulated output spectrum of the current-scaled single-stage fourth-order delta-sigma modulator of Fig. 4.22.

However, the signal swing is different in the current-scaled delta-sigma modulator of figure 4.22 compared with the one of figure 4.11. Unlike in the SI delta-sigma of figure 4.11 where every integrator has the same signal swing, we scale down the signal swing after the first integrator in the SI delta-sigma modulator of figure 4.22. In figure 4.24, we show the simulated amplitude histogram. The input signal frequency is 2 kHz, the input signal amplitude is 6 dB below the feedback current to the first integrator (i.e., -6 dBFS), and the clock frequency is 2.048 MHz. We use 20480 data to plot the histogram.

It is seen from figure 4.24 that the signal swing within the first integrator is almost four times as large as the signal swings within all the other integrators. Therefore, the bias current in the first integrator can be designed 4 times as large as the rest integrators. The dynamic range can be increased without a drastic increase in the power dissipation or chip area. If we use the architecture of figure 4.11 to achieve the same dynamic range, the power dissipation and chip area will be more than two times higher than the one shown in figure 4.22. (Assume that the thermal noise is the major limitation of the dynamic range.)

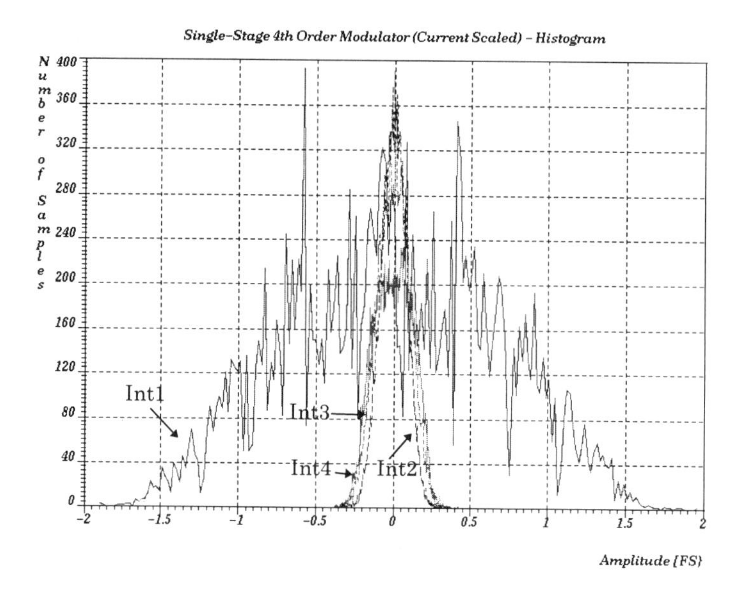

Fig. 4.24. Histogram of the integrator outputs of the current-scaled single-stage fourth-order delta-sigma modulator of figure 4.22.

4.8. Summary

In this chapter, we have discussed various SI delta-sigma modulator architectures. Besides the requirement of the noise shaping function of an SI delta-sigma modulator, it is important to de-couple the settling chain in the successive stages by having delays in the integrators or differentiators. It is also of importance to make sure that the signal swings within all the modulator is identical to facilitate the circuit implementation. Based on these criteria, we have detailed a second-order SI delta-sigma modulator, a two-stage fourth-order SI delta-sigma modulator, a single-stage fourth-order SI delta-sigma modulator, and a system-level chopper-stabilized SI delta-sigma modulator.

To extend the signal bandwidth to higher frequencies, higher-order delta-sigma modulators are needed. With higher-order delta-sigma modulators, we can achieve the same dynamic range with a lower oversampling ratio compared with lower-order delta-sigma modulators. It is usually preferable to use the two-stage fourth-order modulator architecture over the single-stage fourth-order delta-sigma modulator architecture in that the two-stage architecture is more regular and has a larger relative input. Though the single-stage modulator may be able to achieve a higher dynamic range owing to the fact that the matching between digital and analog circuits is not required, the two-stage fourth-order SI delta-sigma modulator is well suited for high speed applications.

By using the system-level chopper stabilization, low frequency noise and/or disturbance introduced at the modulator input can be eliminated. It is effective when we use the first-generation SI circuits to design oversampling A/D converters for audio applications. When the second-generation SI circuits are used, the low frequency noise is automatically reduced due to the inherent correlated double sampling, obviating the need for the chopper stabilization.

Large thermal noise in SI circuits may limit the ultimate dynamic range that SI delta-sigma modulators can achieve. Increasing the power is the only way to increase the dynamic range in a thermal-noise limited SI delta-sigma modulator. Since only the noise at the input sets the limit, we can use different bias current for the first and the other integrators (or differentiators) by scaling down the signal after the first integrator (or differentiator). Thereby we can increase the dynamic range of a thermal-noise limited SI delta-sigma modulator at a moderate cost of power dissipation and chip area consumption. The ratio of the signal swings of (therefore the bias currents for) the first and the other integrators (or differentiators) in a current-scaled modulator is usually around 2~8 due to the practical consideration.

We will present several measured SI delta-sigma modulator chips in Chapter VII after discussing the building blocks and the practical issues in the following two chapters.

Chapter V: Building Blocks for SI Oversampling A/D Converters

5.1. INTRODUCTION

As discussed in Chapter IV, we need integrators, differentiators, current quantizers, and 1-bit D/A converters to construct SI delta-sigma modulators. To control SI delta-sigma modulators, clock generators are needed. In some applications, the input analog signal is a voltage signal, a voltage-to-current (V/I) converter is usually needed before driving an SI oversampling delta-sigma modulator. Due to the advantage of oversampling, a simple low pass filter sometimes suffices as the anti-aliasing filter for oversampling A/D converters. It is of interest to integrate such a filter in the digital CMOS process. To convert the output of an SI delta-sigma modulator into a more convenient digital representation, a digital decimation filter is needed to remove the high-frequency noise and reduce the sampling rate. To open switches for 1.2-V SI circuits, voltage doublers are needed. To collect the output data of a low-voltage oversampling A/D converters into a computer, an on-chip level conversion is also needed. In this chapter, we will discuss all these building blocks.

5.2. SI INTEGRATORS

To construct SI delta-sigma modulators shown in figures 4.3, 4.7, 4.11, 4.19, and 4. 22, we need lossless SI integrators.

Lossless integrator are easy to construct by using the practical SI circuits discussed in Chapter III. Cascading two memory cells (forming a delay line) and feeding back the output of the second memory cell to the input of the first memory cell results in a lossless integrator. Current mirrors are used to output the currents, realizing the scaling factor. In figure 5.1, we show a conceptual fully differential lossless SI integrator.

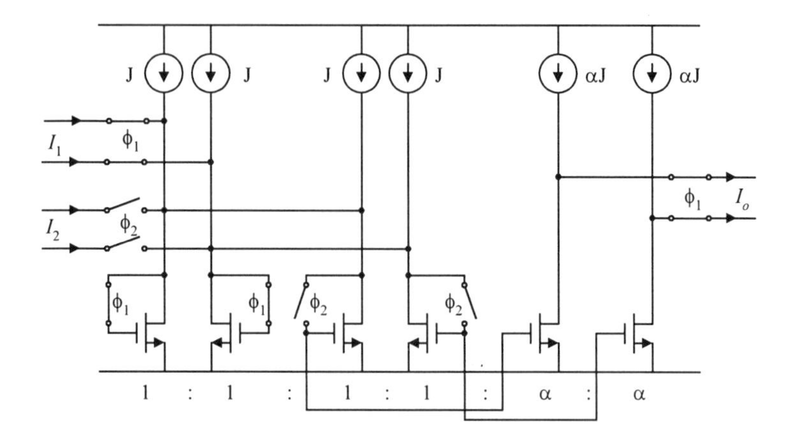

Fig. 5.1. Fully differential lossless SI integrator.

All the input transistors (connected with the switches) have the same dimension, and the dimension of the output transistors is α times as large as that of the input transistors. If the output current I_O of the lossless integrator is sampled on clock phase ϕ_1, we have the signal transfer function given by

$$I_o(z) = \frac{\alpha z^{-1}}{1 - z^{-1}} \cdot I_1(z) - \frac{\alpha z^{-1/2}}{1 - z^{-1}} \cdot I_2(z) \tag{5.1}$$

where $I_o(z)$ is the output current, $I_1(z)$ and $I_2(z)$ are the input currents, all expressed in the z domain, and α is the scaling factor. From the input I_1 to the output I_O there is one clock period delay and from the input I_2 to the output I_O there is only a half clock period delay.

By replacing the memory cells in the lossless integrator of figure 5.1 with the practical SI memory cells discussed in Chapter III, we can get practical lossless SI integrators that can be used in SI delta-sigma modulators.

To use the single-ended memory cell such as the one of figure 3.1, only one branch of figure 5.1 is needed.

To use the memory cell with GGAs such as the ones of figures 3.3 (b) and 3.7, the GGAs and the level shifter can be shared by the two cascaded memory cells. Such an integrator is shown in figure 5.2. The GGAs and the level shifter are shared by the two memory cells. All the bias voltages are not shown in the figure.

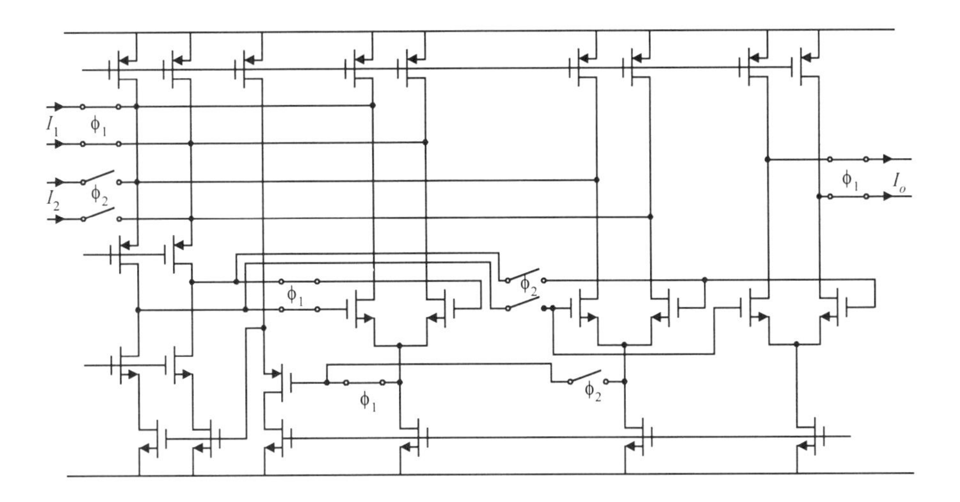

Fig. 5.2. Lossless SI integrator based on the memory cell of Fig. 3.3 (b).

To use memory cells without internal CMFBs such as the ones of figures 3.7, 3.10, 3.16, 3.21, 3.25, and 3.26, at least one CMFF circuit based on the principle illustrated in figure 3.5 or 3.6 must be included not to let the common-mode current accumulate. We just show two examples here.

Shown in figure 5.3 (a) is a lossless SI integrator using the S^2I memory cell of figure 3.21 (b) and shown in figure 5.3 (b) is the CMFF circuit. The bias voltages are not shown in the figure.

In figure 5.3 (b), nodes 1 and 2 are connected to the gates of the coarse memory transistor pair and nodes 3 and 4 are connected to the gates of the fine memory transistor pair, respectively. The transistors are dimensioned in such a way that the output current I_{cm} is equal to the common-mode current of the memory cell that the CMFF circuit is connected to.

(It is also possible to use the current scaling CMFF principle to save power. Detailed discussions have been covered in Chapter 3.4.)

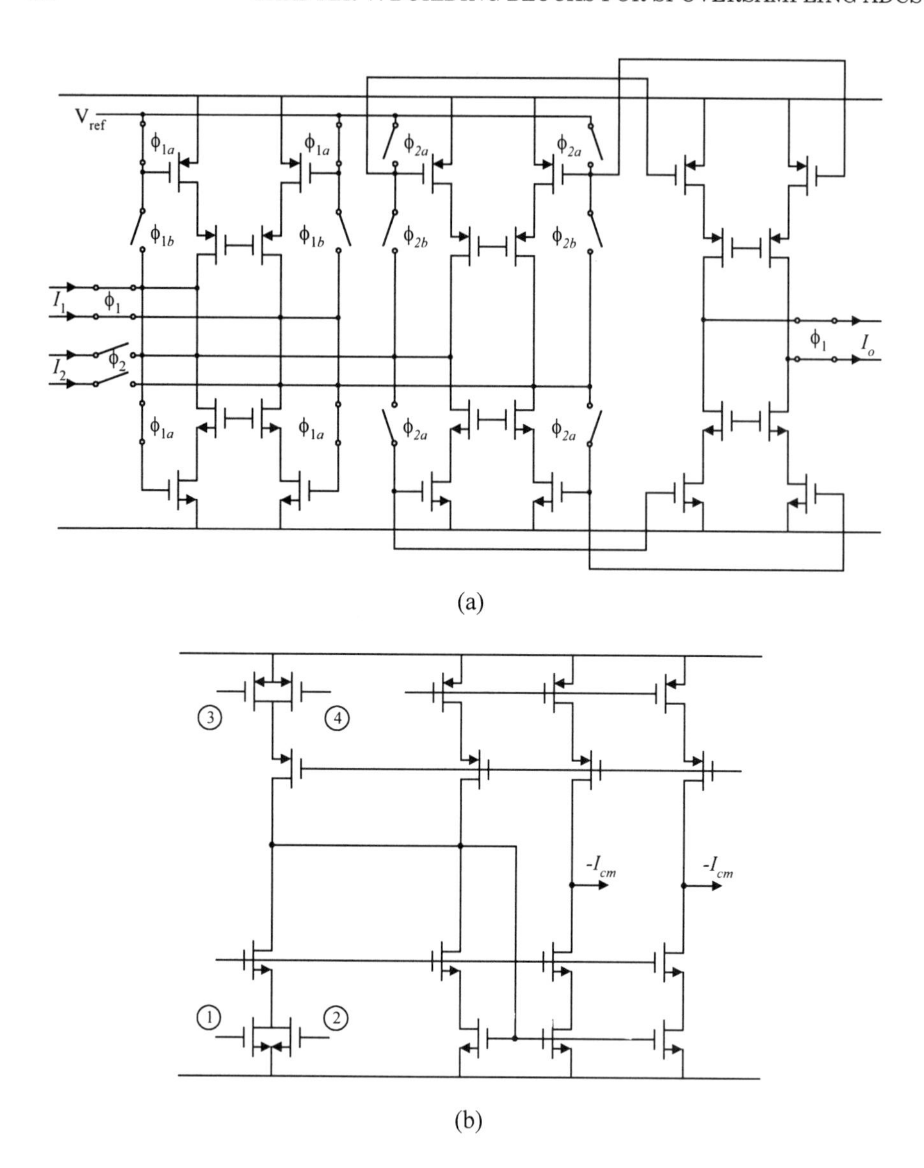

Fig. 5.3. (a) Fully differential lossless S^{2}I integrator using the memory cell of Fig. 3.21 (b), and (b) CMFF circuit.

The two output currents of the CMFF circuit of figure 5.3 (b) are then connected to the differential outputs of the memory cell (i.e., the fully differential inputs of the lossless integrator) during its output clock phase. To avoid feedback due to the CMFF circuit, the output current of the CMFF circuit is only provided on one clock phase. If the CMFF circuit is used for the first memory cell that is controlled by clock phases ϕ_1, ϕ_{1a}, ϕ_{1b}, the output of the CMFF circuit is only connected with the input of the lossless integrator on clock phase ϕ_2. If the CMFF circuit is used for the second memory cell that is controlled by clock phases ϕ_2, ϕ_{2a}, ϕ_{2b}, the output of the CMFF circuit is only connected with the input of the lossless integrator on clock phase ϕ_1.

Another example is shown in figure 5.4. Shown in figure 5.4 (a) is a lossless SI integrator using the ultra low-voltage memory cell of figure 3.16 and shown in figure 5.4 (b) is the CMFF circuit.

In figure 5.4 (b), nodes 1 and 2 are connected to the gates of the memory transistor pair, respectively. All the transistors have the half size of corresponding ones in the memory cell. Therefore, the output currents are equal to the common-mode current of the memory cell but with the direction inverted.

(Notice that it is also possible to use the current scaling CMFF principle to generate the common-mode currents to save power. Detailed discussions have been covered in Chapter 3.4.)

By connecting the output currents of the memory cell (i.e., the input currents of the integrator) and the output currents of the CMFF circuit together, we generate the currents that do not contain any common-mode components. To avoid feedback due to the CMFF circuit, the output currents of the CMFF circuit are only provided on one clock phase. If the CMFF circuit is used for the first memory cell that is controlled by clock phase ϕ_1, the output of the CMFF circuit is only connected with the input of the lossless integrator on clock phase ϕ_2. If the CMFF circuit is used for the second memory cell that is controlled by clock phase ϕ_2, the output of the CMFF circuit is only connected with the input of the lossless integrator on clock phase ϕ_1.

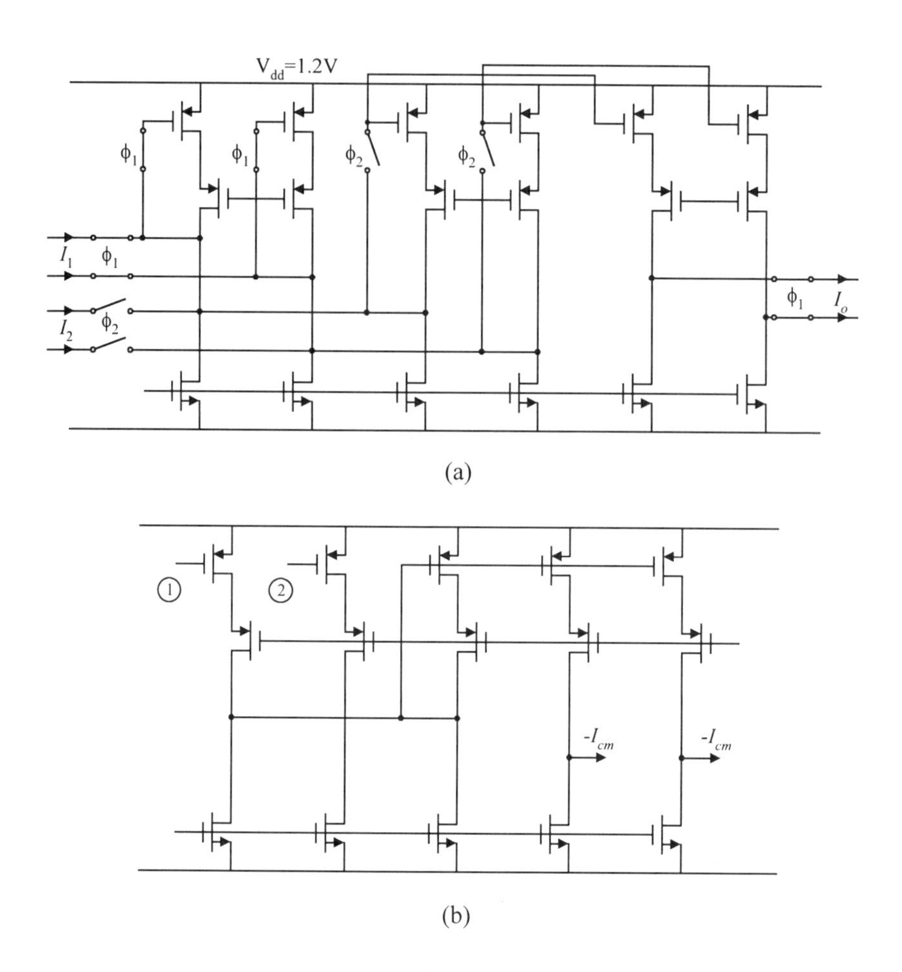

Fig. 5.4. (a) Fully differential lossless SI integrator using the 1.2-V SI memory cell of Fig. 3.16, and (b) CMFF circuit.

5.3. SI DIFFERENTIATORS

To construct the system-level chopper-stabilized delta-sigma modulator of figure 4.15, we need lossless SI differentiators. Lossless differentiators are easy to construct by using the SI circuits discussed in Chapter III. Cascading two memory cells (forming a delay line) and subtracting the output of the second memory cell from the input of the first memory cell results in a lossless differentiator. For fully-differential realizations, the subtraction is realized just by interchanging the currents. Current mirrors are used to output currents, realizing the scaling factor. In figure 5.5, we show a conceptual fully differential lossless SI differentiator.

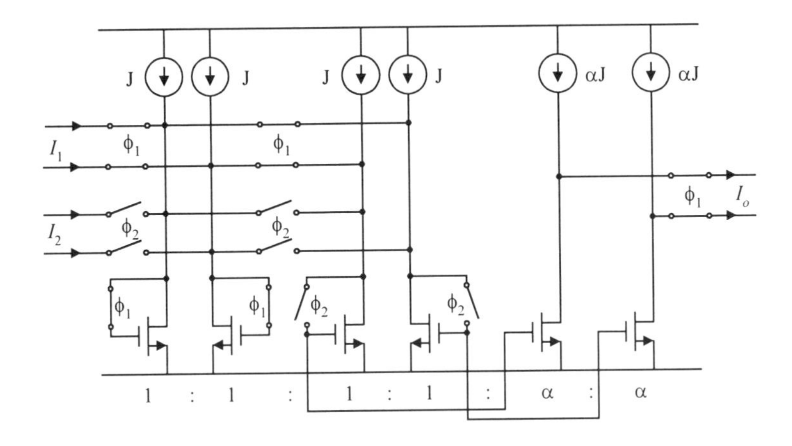

Fig. 5.5. Fully-differential lossless SI differentiator.

All the input transistors (connected with the switches) have the same dimension, and the dimension of the output transistors is α times as large as that of the input transistors. If the output current I_o of the lossless SI differentiator is sampled during clock phase ϕ_1, we have the signal transfer function given by

$$I_o(z) = \frac{\alpha z^{-1}}{1+z^{-1}} \cdot I_1(z) - \frac{\alpha z^{-1/2}}{1+z^{-1}} \cdot I_2(z) \qquad (5.2)$$

where $I_o(z)$ is the output current, $I_1(z)$ and $I_2(z)$ are the input currents, all expressed in the z domain, and α is the scaling factor. From the input I_1 to the output I_o there is one clock period delay and from the input I_2 to the output I_o there is only a half clock period delay.

By replacing the memory cells in the lossless differentiator of figure 5.5 with the practical SI memory cells discussed in Chapter III, we can get practical lossless SI differentiators that can be used in the system-level chopper-stabilized SI delta-sigma modulator.

To use the single-ended memory cells such as the one of figure 3.1, only one branch of figure 5.5 is needed and an extra current mirror is needed to realize the inversion of the feedback current.

To use the memory cells with GGAs such the ones of figures 3.3 (b) and 3.7, the GGAs and the level shifter (only present in the memory cell of figure 3.3 (b)) can be shared by the two cascaded memory cells.

To use memory cells without internal CMFBs such as the ones of figures 3.7, 3.10, 3.16, 3.21, 3.25, and 3.26, at least one CMFF circuit based on the principle illustrated in figure 3.5 or 3.6 must be included not to let the common-mode current accumulate. Here, we just show one example in figure 5.6.

Shown in figure 5.6 (a) is the fully differential lossless SI differentiator using the class-AB SI memory cell of figure 3.7 and shown in figure 5.6 (b) is the CMFF circuit. The bias voltages for the cascode transistors, current source transistors, and GGA transistors are not shown in the figure.

The GGAs are shared by the two memory cells in the fully differential lossless SI differentiator as shown in figure 5.6 (a). One GGA is used in the CMFF circuit to increase the input conductance of the current mirror as shown in figure 5.6 (b). (The CMFF circuit shown in figure 5.6 (b) is detailed in Chapter 3.5.) By connecting the output currents of the memory cell that the CMFF is connected to and the output currents of the CMFF circuit together, we generate the output currents that do not contain any common-mode components. To avoid feedback due to the CMFF circuit, the output current of the CMFF circuit is only provided on one clock phase. If the CMFF circuit is used for the first memory cell that is controlled by clock phase ϕ_1, the output of the CMFF circuit is only connected with the input of the lossless differentiator (i.e., the output of the memory cell that the CMFF is connected to) on clock phase ϕ_2. If the CMFF circuit is used for the second memory cell that is controlled by clock phase ϕ_2, the output of the CMFF circuit is only connected with the input of the lossless differentiator (i.e., the output of the memory cell that the CMFF is connected to) on clock phase ϕ_1.

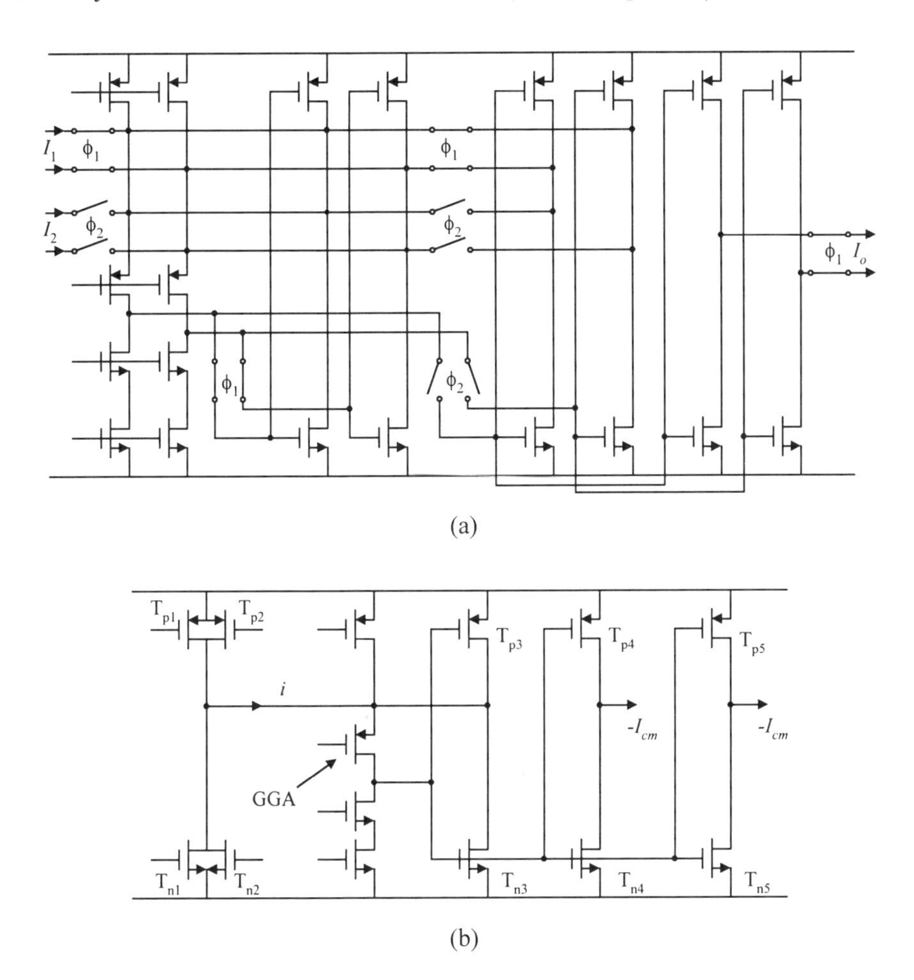

Fig. 5.6. (a) Fully differential lossless SI differentiator using the class-AB SI memory cell of Fig. 3.7, and (b) CMFF circuit.

5.4. CURRENT QUANTIZERS

A current quantizer is also referred to as a current comparator. It detects the current flow direction and generates a voltage output that can be directly amplified to the CMOS logic level. Different current quantizer structures exist. In figure 5.7, we show four different current quantizers.

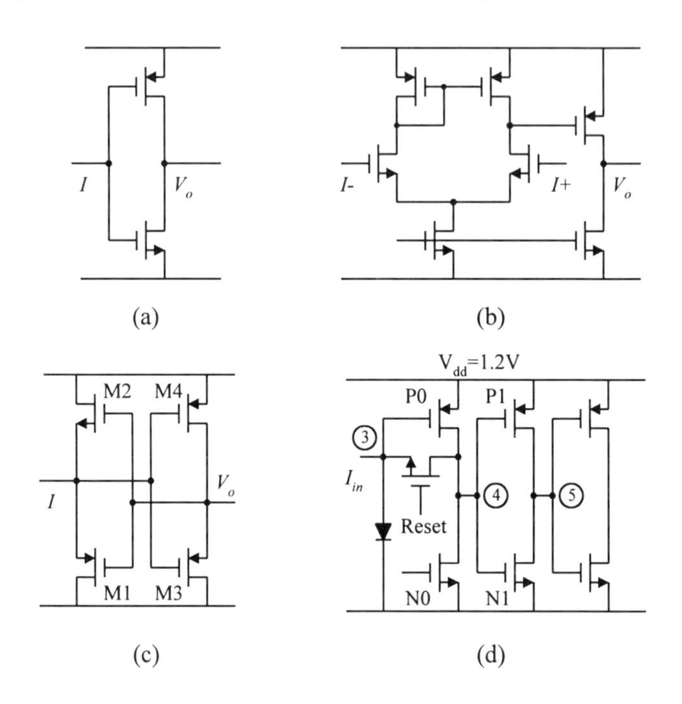

Fig. 5.7. Current quantizers. (a) inverter as a current quantizer; (b) differential voltage comparator as a current quantizer; (c) low-impedance current quantizer; and (d) ultra low-voltage current quantizer.

For the current quantizer of figure 5.7 (a) [27, 43], the input current charges or discharges the input parasitic capacitance and causes the potential at the inverter input to change. For a positive input current I, i.e., the current is forced into the inverter, the voltage at the inverter input ramps up until it hits the upper voltage rail V_{dd} causing the inverter output to go low. For a negative input current I, the current is pulled to ground and the inverter output goes high. Since the input impedance of an inverter is very high, this kind of current quantizers are not good for high speed high resolution applications. To toggle the output, a reasonably large voltage change at the input is needed due to the limited gain of the inverter. It takes time for a small current to charge or discharge the input capacitance. To increase the gain of the inverter, we usually have to increase the transistor size, increasing the input capacitance and therefore decreasing the speed.

To use the current quantizer of figure 5.7 (a), a fully-differential to single-ended converter is needed before driving the current quantizer. In the fully-differential to single-ended converter, we only need a current mirror realizing the inversion of the current direction of one branch and then connect together the two branch currents (one inverted due to the current mirror). The output current of the fully-differential to single-ended converter is the difference of the fully different input currents.

For the current quantizer of figure 5.7 (b) [29, 30], fully differential SI circuits can directly drive it. The currents from the preceding SI circuits charge or discharge the gate parasitic capacitance of the input transistors of the comparator depending on the direction of the currents. When one branch of the differential outputs from the preceding SI circuits charges the gate parasitic capacitance of one input transistor of the current quantizer, the other discharges the gate parasitic capacitance of the other input transistor of the current quantizer. The voltages at the current quantizer's input differential nodes ramp in the opposite direction towards V_{dd} or the ground, respectively. The differential voltage at the current quantizer's inputs thus changes according to the direction of the current flows and the output is produced according to the differential voltage. This quantizer also suffers from the same problem of the current quantizer of figure 5.7 (a) due to the high input impedance. Another serious drawback of this current quantizer is the undefined common-mode voltage for the fully differential current inputs. Unlike in SC applications where the driving circuit has a low impedance and the common-mode voltage is defined by the driving circuit, in SI circuits, the driving circuit has a high impedance that makes the common-mode voltage drift when the current is applied to the high impedance, though the differential voltage due to the input currents is well defined. The drifting of the common-mode voltage at the input may reduce the gain of the comparator and increase the output conductance of the driving circuit (it makes some transistors out of normal operation regions), introducing errors.

For high speed applications, low-impedance current quantizers such as proposed in [56] are needed. The low-impedance current quantizer is shown in figure 5.7 (c) [10, 32, 34]. The drawback is the large quiescent power consumption due to the large gate-source voltages for both N- and P-type transistors in series between the two supply rails.

When the input current I flows into the current quantizer of figure 5.7 (c), all the current must be sunk by the P-type transistor M1, making its gate-source voltage less than the value of its threshold voltage. When the input current I flows from the current quantizer, all the current must be provided by the N-type transistor M2, making its gate-source voltage larger than its threshold voltage. Therefore, the gates of M1 and M2 experience a large potential change when the input current changes the direction. Transistors M3 and M4 act as a voltage amplifier. The small voltage variation at the input due to the input current is amplified. The input conductance of the current quantizer is the transconductance of the input device M1 (or M2) multiplied by the gain of the voltage amplifier consisting of transistors M3 and M4. The input conductance g_{in} is therefore given by

$$g_{in} = g_m \cdot A \tag{5.3}$$

where g_m is the transconductance of the input transistor M1 or M2 (depending on the input current direction) and A is the gain of the voltage amplifier consisting of transistors M3 and M4. Due to the low input impedance, the potential change at the input is small.

The speed of the current quantizer of figure 5.7 (c) is determined by the input impedance, the input parasitic capacitance, and the response time of the voltage amplifier consisting of M3 and M4. After optimization, a settling time of 10 ns can be achieved with a current step input of 0.5 μA in a standard 3-V 0.8-μm CMOS process [17].

To use the current quantizer of figure 5.7 (c), a fully-differential to single-ended converter is needed before driving the current quantizer.

To achieve a high speed and high resolution current comparison with the ultra low-supply voltage, we can use a resettable current quantizer shown in figure 5.7 (d) [11].

When Reset is high, transistor P0 is diode connected and transistor N0 provides the bias current. We design the DC voltage at node 4 close to the corresponding DC voltage of the driving circuit. Transistors N1 and P1 form an inverter and are dimensioned to have high speed. A minimum sized inverter is used as the load. Due to the very low supply voltage (less than twice the threshold voltage), only the branch consisting of transistors N0 and P0 conducts a DC current. This DC current can be designed very small (e.g., $< 2\ \mu A$). Therefore, the current quantizer of figure 5.7 (d) does not consume much static power.

When Reset is high and there is a current I_{in} flowing into the current quantizer, the potential of nodes 3 and 4 changes. When Reset goes low, the gate of P0 is isolated from its drain and the gate voltage changes the potential at node 4. At the same time, the gate voltage keeps changing due to the input current. This makes the potential change at node 4 even faster. Therefore, a fast and accurate comparison can be accomplished.

To reduce the settling time when Reset goes high, we use a diode at the input to limit the potential.

In figure 5.8, we show the simulation result of the current quantizer of figure 5.7 (d). The simulator is HSPICE and the process is the 0.8-μm digital CMOS process using which the 1.2-V memory circuit of figure 3.16 is designed. A current mirror with the same architecture as the SI memory cell of figure 3.16 is used to drive the current quantizer to emulate the real situation.

With an input current 50 nA, a comparison time of less than 0.25 μs is achieved as seen in figure 5.8. When the input current increases to 1 μA, the comparison time is less than 15 ns. And the power consumption is less than 2.4 μW.

To use the current quantizer of figure 5.7 (d) for fully differential SI circuits, a fully-differential to single-ended converter is needed before driving the current quantizer.

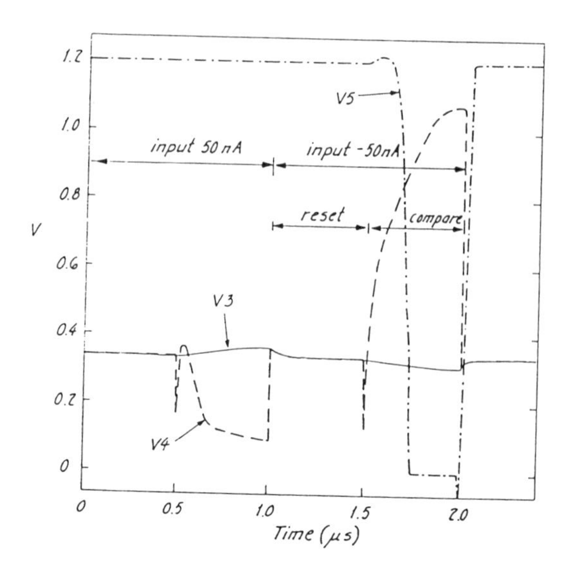

Fig. 5.8. Simulated response of the ultra low-voltage current quantizer of Fig. 5.7 (d).

5.5. One-Bit D/A Converters

The 1-bit D/A converter in an SI delta-sigma modulator is a current source controlled by the current quantizer output. The output current direction changes according the digital input that is the output from the current quantizer in the delta-sigma modulator. The current values can be changed simply by changing the transistor sizes of the current sources. The type of current sources can be chosen to be similar to the memory cell structure. In figure 5.9, we show the three different types of 1-bit D/A converters.

The two output currents change the flow direction according to the digital input C (i.e., the current quantizer output). The difference in the current values of the positive and negative currents only introduces a DC offset. The small DC offset usually does not present any serious problem for 1-bit delta-sigma modulators [41].

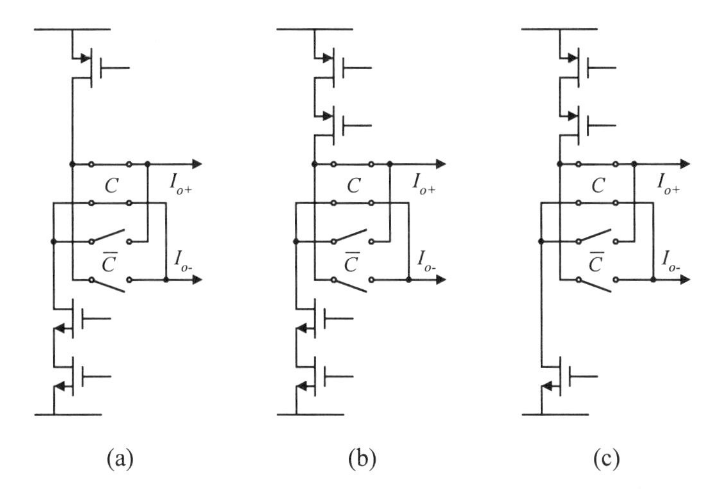

Fig. 5.9. 1-bit D/A converters.

The 1-bit D/A converter of figure 5.9 (a) can be used in the SI delta-sigma modulators using memory cells of figures 3.3 and 3.7. The output potential of the 1-bit D/A converter is fixed by the GGAs in the integrators where the output current of the 1-bit D/A converter goes.

The 1-bit D/A converter of figure 5.9 (b) uses cascode current sources. Small variation of the potential at the output does not change the current value much due to the cascoding. It can be used in the SI delta-sigma modulators using the memory cells of figures 3.10 and 3.21 (b).

The 1-bit D/A converter of figure 5.9 (c) can be used in the SI delta-sigma modulator using the memory cell of figure 3.16 [11], since their architectures are similar.

5.6. Clock Generators

For a sampled-data system such as an SI system, we usually need clock generators to control the system. For SI circuits, we need non-overlapping clock generators just as for SC circuits. Such a non-overlapping clock generator is shown in figure 5.10 [57]. Non-overlapping clock phases must be used for voltage-sampling switches (switches that sample the gate voltages) in SI circuits.

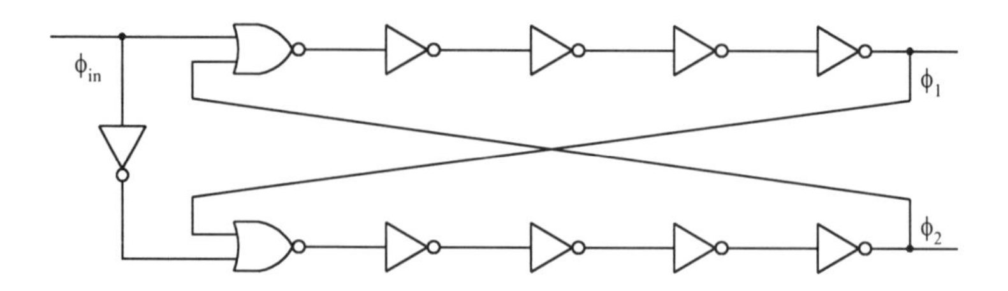

Fig. 5.10. Non-overlapping clock generator.

Non-overlapping clock phases are generated by cross-coupled delay chains as shown in figure 5.10. The non-overlapping interval is controlled by the number of inverters in the delay chains.

In order to reduce the glitch in the second-generation SI circuits, overlapping clock phases can be used for current-steering switches (switches that direct the currents). This will be discussed in more detail in the next chapter. Such an overlapping clock generator is shown in figure 5.11. The outputs are the inversion of the outputs of the non-overlapping clock generator. The timing relationship between the non-overlapping clock generator and overlapping clock generator can be adjusted by inserting inverters.

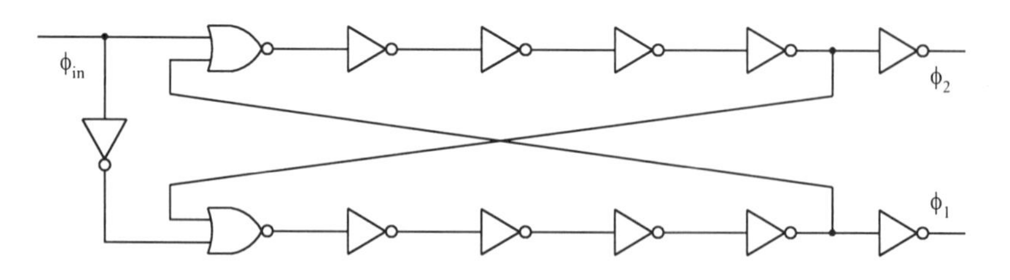

Fig. 5.11. Overlapping clock generator.

5.7. VOLTAGE-TO-CURRENT CONVERTERS

If the input signals are voltages, we need voltage-to-current (V/I) converters to drive SI oversampling A/D converters. They are different ways to construct the V/I converters. We can use commercial operational amplifiers and BJTs [58] or even current conveyors [59] to construct off-chip V/I converters. We can also construct V/I converters on chip. For many applications, it is preferably to use on-chip V/I converters to reduce the number of components on the printed circuit board (PCB).

By using high-gain and low-noise operational amplifiers, very linear off-chip V/I converters with low noise can be designed. The speed is usually not very high (depending on the components). Using current conveyors, we can construct high speed off-chip V/I converters. But they are usually noisy.

There are several ways to construct on-chip V/I converters in CMOS as well. One way is to explore the intrinsic properties (i.e., without feedback) of MOS transistors and improve the linearity by using techniques such as the adaptive biasing, class-AB, source degeneration, and current differencing techniques, etc. [60]. They are relatively complex. Another simpler way is to create a virtual ground and derive the current through a resistor connecting the virtual ground and the voltage source. Such a V/I converter is shown in figure 5.12.

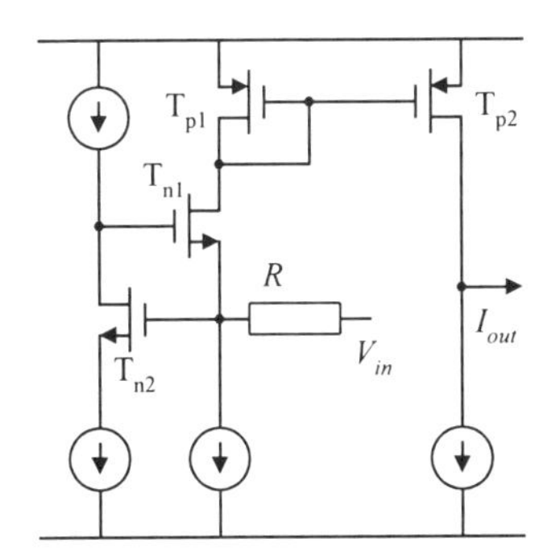

Fig. 5.12. V/I converter in CMOS.

The input impedance at the source of transistor T_{n1} is reduced by the gain of transistor T_{n2}. The equivalent input impedance is given by

$$R_{in} \approx \frac{1}{g_{mn1}} \cdot \frac{1}{A_{n2}} \approx \frac{g_{dsn2}}{g_{mn1} \cdot g_{mn2}} \tag{5.4}$$

where g_{mn1} is the transconductance of transistor T_{n1}, A_{n2} is the gain of transistor T_{n2}, g_{mn2} is the transconductance of transistor T_{n2}, and g_{dsn2} is the output conductance of transistor T_{n2}. Due to the gain of transistor T_{n2}, the input impedance seen at the source of transistor T_{n1} is very small (less than 1 ohm is easily achievable). Therefore, a virtual ground is created at the source of transistor T_{n1}. The current is converted and mirrored out via transistors T_{p1} and T_{p2}. Suppose the same size and bias for transistors T_{p1} and T_{p2}, we have the output current I_{out} given by

$$I_{out} = \frac{V_{in}}{R + R_{in}} \approx \frac{V_{in}}{R} \tag{5.5}$$

where R is the resistance shown in figure 5.12.

From equation (5.4) it is seen that the input impedance R_{in} at the virtual ground is signal dependent due to the strong dependency of the transconductance g_{mn1} on the input current. To reduce the error of the V/I converter, we usually require that $R >> \Delta R_{in}$ (where ΔR_{in} represents the variation of the input impedance within the whole input signal range), or simply $R >> R_{in}$. If we require the error in the V/I converter less than -60 dB, the input impedance at the virtual ground must be $R_{in} < 10^{-3} \cdot R$.

If we put a pad at the source of transistor T_{n1} (i.e., resistor R is external), the large capacitance at the pad does not limits the speed since the impedance is low. If we put a pad at the input terminal (i.e., resistor R is internal), since the voltage source has a low impedance, the large capacitance at the pad does not present any problem either. Therefore, the V/I converter shown in figure 5.12 is very good for high speed operations.

For most modern digital CMOS processes, good resistors are not available, and the poly gate is silicidated. The sheet resistance of the silicidated poly gate is usually around 1~2 ohm/square. It may therefore not be a good choice to use internal resistors created by the low-ohmic poly layer. When an external resistor is used, special caution should be taken to avoid oscillation, especially at high frequencies. The risk of oscillation is due to the existence of the large inductance (bondwire) and the frequency dependency of the input impedance of the V/I converter. Sometimes it is rewarding to split the resistor R into two parts, a small part on-chip, and a large part external. Then the influence of the bondwire inductance can be reduced.

For measurement purposes, an AC coupled input can be used for the V/I converter of figure 5.12 to avoid the large offset current if the DC voltage at the virtual ground and that at the input terminal is not matched. For some real applications where the DC coupling is requested, efforts must be taken to design the DC voltage of the virtual ground of the V/I converter to avoid excess off-set currents that may saturate the following SI circuit. We show two methods here.

(Notice that in most real applications, the DC voltage of the AC input signal or the common-mode voltage (for fully differential inputs) are explicitly provided.)

Shown in figure 5.13 is the method to cancel the DC offset current due to the mismatch between the DC voltage of the input AC signal and the DC voltage of the virtual ground.

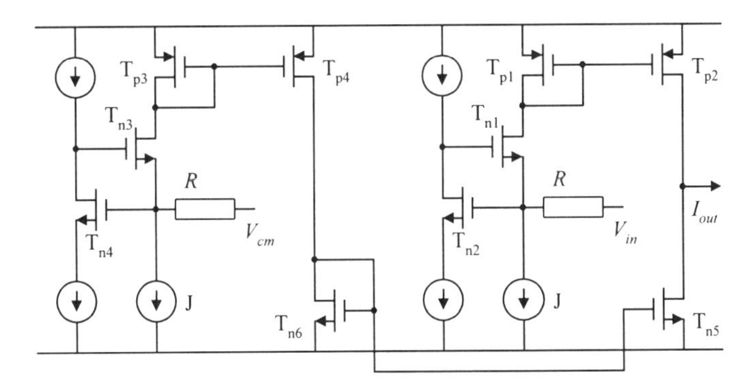

Fig. 5.13. Method to cancel the DC offset current by adapting the bias current.

The main idea is to cancel the DC offset in the output current. Instead of using an independent bias current source for the output transistor T_{p2}, we use a bias current source that is dependent on the DC voltage of the input AC signal (for single-ended circuits), or the common-mode input voltage V_{cm} (for fully differential circuits). The difference between V_{cm} and the DC voltage of the virtual ground at the source of transistor T_{n3} is used to generate a current and this current is mirrored out as a bias current for the output transistor T_{p2}. If all the components are piece-wisely matched, we have

$$I_{out} = I_{dp2} - I_{dn5} \approx J + \frac{1}{R}\left(V_{in} + V_{cm} - V_0\right) - J - \frac{1}{R}\left(V_{cm} - V_0\right) = \frac{V_{in}}{R} \quad (5.6)$$

where I_{dp2} and I_{dn5} are the drain current of transistor T_{p2} and T_{n5} respectively, J is the quiescent bias current for transistor T_{p1} (and T_{p3}), V_{in} is the input AC voltage, V_{cm} is the DC voltage of the input AC signal (for single-ended circuits) or the common-mode input voltage (for fully differential circuits), and V_0 is the DC voltage at the virtual ground.

It is seen from equation (5.6), the output current is free from the DC offset current, which is given by

$$I_{os} = \frac{1}{R}\left(V_{cm} - V_0\right) \quad (5.7)$$

In reality, the DC offset current still exists due to the mismatch between the corresponding components. However, this DC offset current is usually very small and will not saturate the following SI circuits.

To save the power dissipation, the circuit block shown in the left-hand side of figure 5.13 does not have to have the same bias current condition as the circuit block shown in the right-hand side. For example, we can use the left-hand side circuit to generate a current equal to 1/4 of the offset current I_{os} given by equation (5.7) and make transistor T_{n5} four times as large as transistor T_{n6}. Then equation (5.6) still holds. However, the current dissipation by the left-hand side circuit can be reduced approximately by a factor of 4.

Another simple way to get rid of the DC offset current is to directly control the DC voltage of the virtual ground. This is illustrated in figure 5.14.

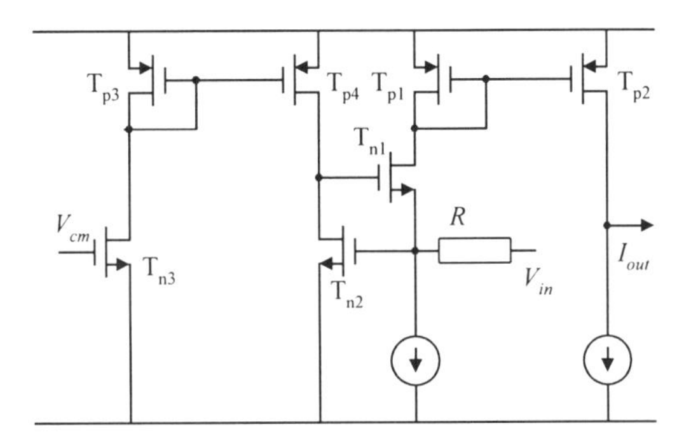

Fig. 5. 14. Method to cancel the DC offset current by directly controlling the DC voltage of the virtual ground.

Due to the current mirror formed by transistors T_{p3} and T_{p4}, the drain currents of T_{n2} and T_{n3} are forced to be equal. Therefore, they must have the same gate-source voltage. Since the source voltages are the same, the gate voltage of transistor T_{n2} (the DC voltage of the virtual ground) is equal to the gate voltage of transistor T_{n3} (V_{cm}). Therefore, the output current is given by

$$I_{out} = \frac{1}{R}\left(V_{in} + V_{cm} - V_0\right) \approx \frac{1}{R}\left(V_{in} + V_{cm} - V_{cm}\right) = \frac{V_{in}}{R} \tag{5.8}$$

where V_{in} is the input AC voltage, V_{cm} is the DC voltage of input AC signal (for single-ended circuits) or the common-mode input voltage (for fully differential circuits), and V_0 is the DC voltage of the virtual ground.

The channel length modulation, as well as the matching of the transistors, introduces errors in the matching of the gate voltages of T_{n2} and T_{n3}, introducing an off-set current. However, this small DC offset current is usually very small and will not saturate the following SI circuits.

To save the power, the current mirror consisting of T_{p3} and T_{p4} can be ratioed so that the current dissipation in T_{p3} is smaller than that of T_{p4}. In this case, transistors T_{n3} and T_{n2} need to be ratioed proportionally as well.

All the above discussions are conceptual. Real designs should use the current mirror structures similar to the SI circuits, e.g., cascode current mirrors, or current mirrors with GGAs, etc.

5.8. Low-Cost On-Chip Filters for SI Oversampling A/D Converters

For real applications, only an A/D converter does not suffice. We need anti-aliasing filters. One of the great advantages of oversampling A/D converters is that the requirement on the anti-aliasing filters is relaxed. Depending on the oversampling ratio and the application environment, one or two-pole filters may suffice for the anti-aliasing purpose. Obviously, it is of great advantage to integrate the anti-aliasing filter on the same chip with the oversampling A/D converter to reduce the cost.

In a digital CMOS process, neither resistors nor linear capacitors are available. Though it is possible to utilize the gate poly as resistors, the sheet resistance is very small (1~2 ohm/square) and has large variation (as large as 100~200%) in a sub-micro CMOS process. Therefore, it is better to use active components, i.e., transistors to realize the resistance. Notice that Nwell (or Pwell) resistors have a sheet resistance around 1 kohm/square and can be used for some applications. They are usually noisy due to the closeness to the substrate and quite nonlinear due to the PN junction formed by the well and substrate.

Though it is possible to utilize the single poly layer and metal layers to realize a linear capacitor to design transconductance-C filters [61], the sheet capacitance is very small in a sub-micron CMOS process. Therefore, it is better to utilize the gate capacitance of MOS transistors which has a much larger sheet capacitance.

A. Current mirror as a low-pass filter

Let's first take a look at a current mirror shown in figure 5.15. A capacitor C_0 is added at the gate in the current mirror as shown in the figure.

The capacitor C_0 can be realized by a gate capacitor on chip or realized by an off-chip capacitor if the cut-off frequency of the filter is required to be very low. By properly dimensioning the sizes of transistors M0 and M1 and their associated bias currents, a scaling factor can also be realized within the current mirror. The current mirror can be also viewed as a single-pole filter and the pole is created by the capacitance C_0 and the transconductance of the input transistor M0.

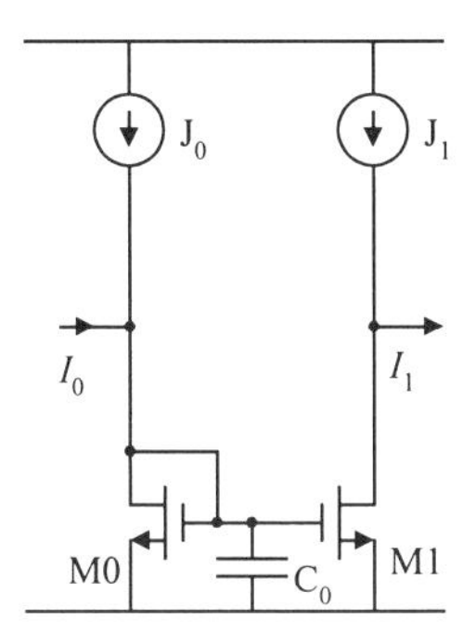

Fig. 5.15. Current mirror as a single-pole low-pass filter.

The pole frequency of the single-pole filter shown in figure 5.15 is given by

$$f_0 = \frac{1}{2\pi} \cdot \frac{g_{m0}}{C_0 + C_{p0}} \tag{5.9}$$

where g_{m0} is the transconductance of the diode-connected transistor M0 and C_{p0} represents all the parasitic at the gate of transistor M0.

If the capacitance changes with the input signal (nonlinear behavior), the pole frequency changes and therefore distortions occur. Hence, the nonlinearity in the capacitance introduces errors in the output current. Though the gate capacitance is highly nonlinear across the whole operation region, in the current mirror configuration as shown in figure 5.15, the gate voltage change is quite limited, making the transistors operate in a well specified region all the time. Therefore, the gate capacitance does not vary dramatically and the linearity is acceptable. (In traditional voltage-mode circuits, the voltage across a capacitor can change between the power supply rails, making the gate capacitance vary dramatically.) If a very low cut-off frequency is needed, the large capacitance can be realized by bond pads or even external capacitors. When pad capacitors or external capacitors are used, the linearity can usually be guaranteed.

However, the pole frequency is also dependent on the transconductance. The transconductance g_m of a transistor is given by

$$g_m = \sqrt{2\mu_n C_{ox} \cdot \frac{W}{L} \cdot I_D} \tag{5.10}$$

where μ_n is the channel charge mobility, C_{ox} is the gate capacitance per square, W/L is the transistor size, and I_D is the drain current. When the drain current in transistor M0 changes accommodating the input current I_0, the transconductance g_{m0} changes, making the pole frequency change.

In figure 5.16, we show the SPICE simulation results when the input current changes between $\pm 0.5 \cdot J_0$, where J_0 is the bias current for the input transistor M0. Cascode current mirrors and cascode current sources are used, and the capacitor C_0 is realized by NMOS transistors.

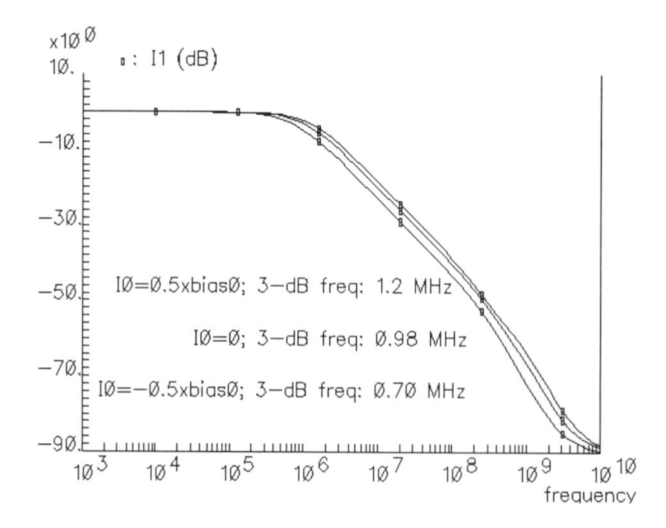

Fig. 5.16. Spice simulation results of Fig. 5.15. Cascode current mirrors and cascode current sources are used, and the capacitor C_0 is realized by NMOS transistors.

It is seen in figure 5.16 that the circuit of figure 5.15 is a single-pole system, having a 20 dB/dec frequency roll-off. And the change in the 3-dB frequency is well in line with the prediction given by equations (5.9) and (5.10).

The change in the pole frequency does not only vary the cut-off frequency but also introduces distortions when the input signal frequency approaches the cut-off frequency in that the input with a different amplitude experiences a different attenuation.

For the same circuit used in simulating the frequency response shown in figure 5.16, the simulated total harmonic distortion is about - 50 dBc when the input is a 100 kHz sinusoid with an amplitude equal to one-fourth of the bias current. When the input frequency decreases to 10 kHz, the total harmonic distortion is less than - 70 dBc. When the input frequency is larger than the cut-off frequency, the harmonic distortions are attenuated by the filter itself.

Obviously, to make the pole frequency well-defined, we need to make the change in the drain current as small as possible. One way to do so is to limit the input current compared to the bias current, thereby limiting the variation in the transconductance of the transistor. This is very power consuming. We will see in the following sections that the proper cascading realizing higher-order filters can reduce the variation in the pole frequencies.

B. Cascading of low-pass filters

A single-pole system only gives a 20-dB/dec roll-off. In many applications, a sharper cut-off is needed. Cascading two single-pole systems realizes a two-pole system having a 40-dB/dec roll-off. Sharper cut-off can be realized by cascading more stages.

There are two possibilities of cascading as shown in figure 5.17.

The use of the cascading shown in figure 5.17 (a) results in a lower power consumption due to the use of the P-type branch. However, it has negative influence on the variation of the pole frequencies. Suppose that input current I_0 is positive, then the drain current in M0 increases making its transconductance increase. Therefore, the pole frequency determined by

g_{m0}/C_0 will increase. At the same time, the drain current in M2 increases as well making its transconductance increase. Therefore, the pole frequency determined by g_{m2}/C_1 will increase as well. The combined effect is that the pole frequencies vary more rapidly as the input current varies.

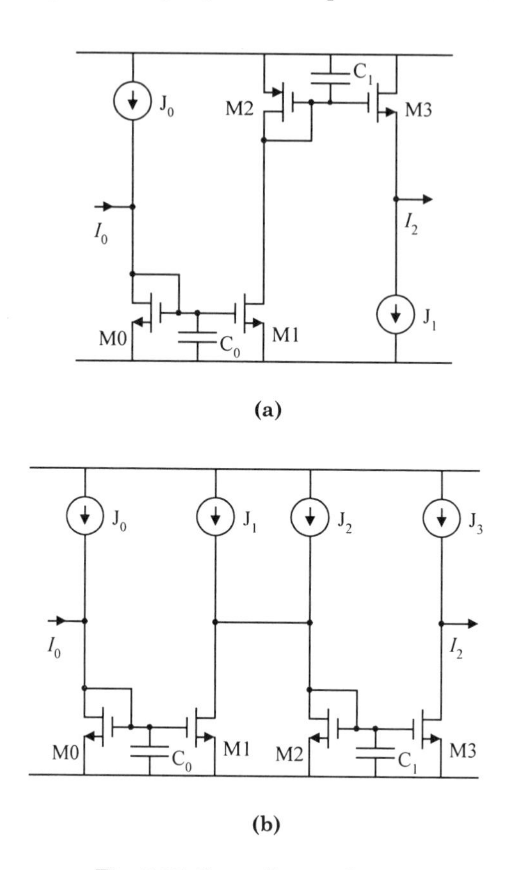

Fig. 5.17. Cascading techniques.

The cascading technique shown in figure 5.17 (b) results in more power consumption due to the extra N-type branch. However, it has a big advantage stabilizing the pole frequencies. Suppose that input current I_0 is positive, then the drain current in M0 increases making its transconductance increase. Therefore, the pole frequency determined by g_{m0}/C_0 will increase. At the same time, the drain current in M2 decreases making its transconductance decrease. Therefore, the pole frequency determined by g_{m2}/C_1 will decrease. The combined effect is that the variations in the two pole frequencies tend to reduce the total variation.

In figure 5.18, we show the SPICE simulation results of the two-pole filter of figure 5.17 (b) when the input current changes between $\pm 0.5 \cdot J_0$, where J_0 is the bias current for the input transistor M0. Cascode current mirrors and cascode current sources are used, and the capacitors C_0 and C_1 are realized by NMOS transistors.

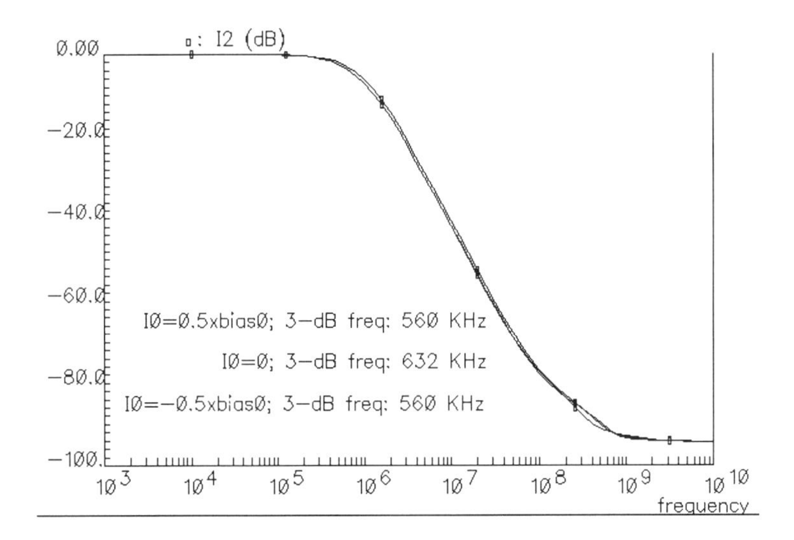

Fig. 5.18. Spice simulation results of Fig. 5.17 (b). Cascode current mirrors and cascode current sources are used, and the capacitors are realized by NMOS transistors.

It is seen from figure 5.18 that the circuit of figure 5.17 (b) is a two-pole system, having a 40-dB/dec frequency roll-off. And the variation in the 3-dB frequency is reduced considerably, compared to the corresponding single-pole system.

The cascading technique shown in figure 5.17 (b) does not only reduce the variation in the cut-off frequency, but also the distortion. For the same circuit used in simulating the frequency response shown in figure 5.18, the simulated total harmonic distortion is less than - 60 dBc when the input is a 100 kHz sinusoid with an amplitude equal to one-fourth of the bias current. When the input frequency decreases to 10 kHz, the total harmonic distortion is less than - 80 dBc. When the input frequency is larger than the cut-off frequency, the harmonic distortions are attenuated by the filter itself. From the simulation results, we see that the total harmonic distortion of the two-pole system of figure 5.17 (b) is considerably smaller than the corresponding single-pole system of figure 5.15.

C. Current mirrors with a reduced bandwidth

In a digital CMOS process, the Nwell (in P-substrate) or Pwell (in N-substrate) having a sheet resistance around 1 kohm/square can be used as resistors, though they have a large variation and poor linearity and are noisy. By introducing such a resistor in a current mirror, we can also reduce the bandwidth to meet the cut-off frequency requirement of a certain filtering. Such a configuration is shown in figure 5.19 [62].

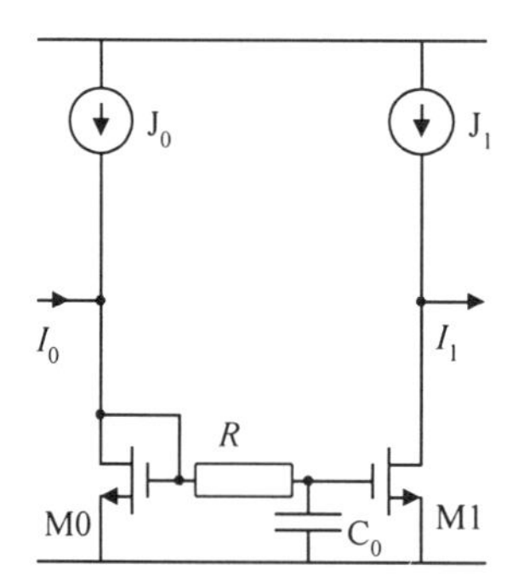

Fig. 5.19. Current mirror with a reduced bandwidth.

The capacitor C_0 is realized by an MOS transistor and the resistor R is realized by the Nwell (in a P-substrate process). If R is much larger than the inverse of the transconductance of transistor M0 and the capacitor C_0 is much larger than the gate capacitance of transistor M0 and M1, the pole frequency can be approximated by

$$f_0 = \frac{1}{2\pi} \cdot \frac{1}{RC_0} \tag{5.11}$$

With the introduction of the resistance R, the cut-off frequency can be reduced. The drawback is the noise due to the noisy resistor at the gate.

D. Combining a V/I converter and a low-pass filter

The on-chip V/I converters discussed in Chapter 5.7 usually have a wide bandwidth. We can easily combine the V/I converter and filter to realize a complete analog interface for SI oversampling A/D converters without needing off-chip components. Such an example is shown in figure 5.20.

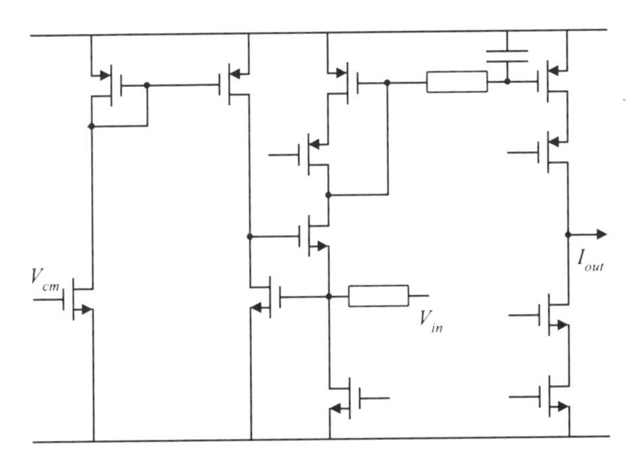

Fig. 5.20. Complete analog interface combining a V/I converter and low pass filter.

The complete analog interface shown in figure 5.20 consists of a V/I converter of figure 5.14 with a controllable DC potential of the virtual ground and a single-pole filter of figure 5.19. Cascode transistors are used for the current mirrors and current sources. All the resistors are realized by using the Nwell and the capacitors are realized by PMOS gate capacitors.

The circuit of figure 5.20 was implemented in Ericsson's in-house 0.6-µm digital CMOS process. In figure 5.21, we show the measured frequency response. The measured cut-off frequency (-3-dB) is 100 kHz, and the attenuation at 2 MHz is 25 dB. It verifies that the circuit of figure 5.20 realizes a single-pole low pass filter.

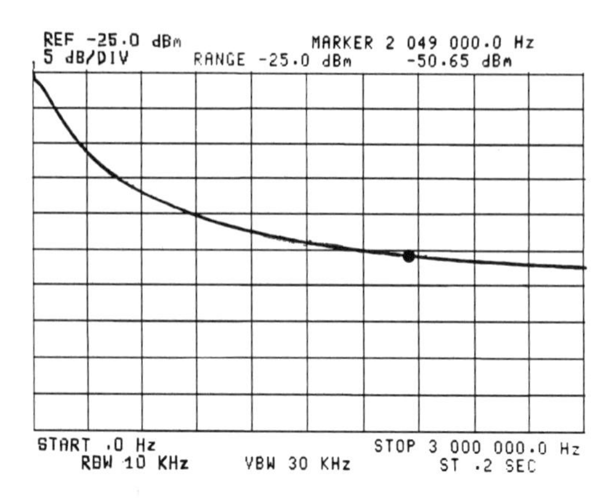

Fig. 5.21. Measured frequency response of the complete analog interface of Fig. 20.

In Table 5.1, we summarize the performance of the complete analog interface of figure 5.20.

Table 5.1. Performance summary of the circuit of figure 5.20.

Supply voltage	3.3 V
Power dissipation (including bias)	7.5 mW
Active chip area (including bias)	800 μm x 150 μm
Cut-off frequency	100 kHz
Attenuation at 2 MHz	25 dB
THD (input: 2 kHz, $I_{bias}/2$)	- 40 dBc
THD (input: 2 kHz, $I_{bias}/4$)	- 46 dBc

5.9. DIGITAL DECIMATION FILTERS

An oversampling A/D converter consists of a delta-sigma modulator and a digital decimation filter. The delta-sigma modulator modulates the analog input to a simple digital code, usually single-bit words, at a frequency much higher than the Nyquist rate. The digital decimation filter smoothes the output of the delta-sigma modulator and reduces the data rate from the oversampling rate to the Nyquist rate. It attenuates the quantization noise, interference, and high-frequency components of the signal before they could alias into the signal band when the code is re-sampled at the Nyquist rate.

Due to practical reasons, it is usually customary to use an appropriate comb (or sinc) filter at the beginning to decimate the modulator output from the oversampling frequency to four times the Nyquist frequency. Then an FIR filter is used to further decimate by a factor of four [41]. The FIR filter is relatively easy to design since the data rate is reduced. The more difficult part is the design of the comb filter since the input data rate is very high.

A. Recursive algorithm of comb filters

The transfer function of a comb filter in the z-domain is given by

$$H(z) = \left[\sum_{i=0}^{N-1} z^{-i} \right]^k = \left[\frac{1 - z^{-N}}{1 - z^{-1}} \right]^k \tag{5.12}$$

where N is the decimation ratio and k indicates the number of cascaded stages. A k-th order comb filter is usually used for (k - 1)-th order delta-sigma modulators.

Notice that sometimes a scaling factor $1/N^k$ is included in the filter function in order to make the gain unity.

An obvious attempt to realize the comb filter is to use a ROM to store all the coefficients and to perform a direct convolution as an ordinary FIR filter. When N is very large which is usually true, this realization becomes unfavorable due to the storing of all the coefficients. In [63], a direct convolution realization of a comb filter without the storing of the coefficients was proposed for oversampling A/D converters. The coefficients were not stored in a ROM but generated by a complicated up/down counter. The regularity was poor. The complexity increases dramatically when N and/or k increase. The most favorable structure has been the one based on the recursive algorithm with an IIR filter followed by an FIR filter [64-66] as shown in figure 5.22. The switch in the figure indicates the reduction of the sampling rate by a factor of N.

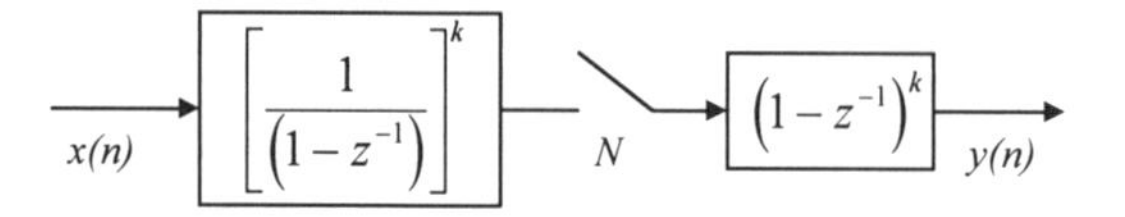

Fig. 5.22. Recursive algorithm for the comb filter.

In order to avoid instability problems, a special modulo arithmetic [64] is needed. The problems are the power consumption and the frequency limitation. The input to the comb filter is from the delta-sigma modulator, usually having a short word length, say m bits, and a high data rate, i.e., the oversampling rate. The data word length inside the IIR part must be at least $m + k\log_2 N$ bits to maintain the accuracy [64]. Since N and k are usually very large, the IIR part has to operate at a very high frequency (the oversampling frequency of the delta-sigma modulator) and have a long word length. Therefore the IIR part limits the applicability of the recursive structure to very high-frequency oversampling A/D converters. The power consumption is not favorable since a lot of calculation is performed using the high frequency clock.

B. Non-recursive algorithm of comb filters

Usually, the decimation ratio N is chosen to be M-th power of 2. Then from equation (5.12), we have

$$H(z) = \left[\sum_{i=0}^{N-1} z^{-i}\right]^k = \left[\sum_{i=0}^{2^M-1} z^{-i}\right]^k$$
$$= \prod_{i=0}^{M-1}\left(1+z^{-2^i}\right)^k \tag{5.13}$$

Thus, by applying the commutative rule [67], a new comb filter algorithm results. It is shown in figure 5.23. The switches in the figure indicate the reduction in the sampling rates by a factor of 2.

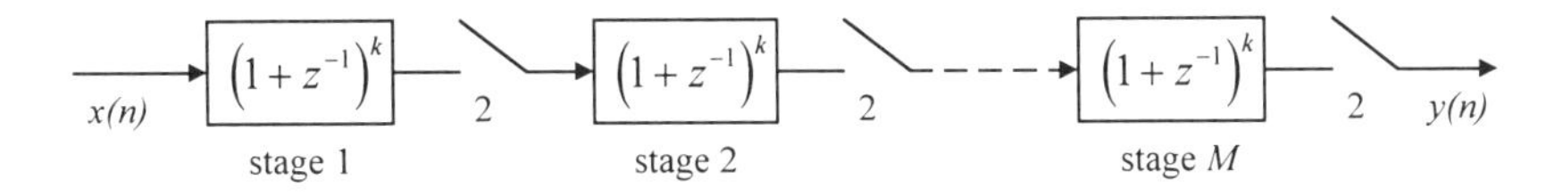

Fig. 5.23. Non-recursive algorithm of comb filters.

This is a non-recursive algorithm. Every stage has the same low-order FIR filter but with a different sampling rate. Thus, this results in a highly regular structure that makes the design and layout easier. Furthermore, reducing the sampling rates as early as possible helps to reduce the workload and thus the power consumption.

The input $x(n)$ is from an oversampling delta-sigma modulator. Its word length W_d is assumed to be m bits. The word length increases through every stage by k bits and the data rate decreases through every stage by a factor of 2 starting from the oversampling rate f_{os}. In Table 5.2, we summarize the observations.

Table 5.2. Data rate and word length of stage i in figure 5.23. $1 \le i \le M$

	Input	Output
Data rate	$f_{os} / 2^{i-1}$	$f_{os} / 2^i$
Word length	$(i-1) \cdot k + m$	$i \cdot k + m$

Referring to figure 5.22, the IIR part is k accumulators in cascade. Although the output sampling rate of the IIR part is f_{os}/N, the sampling rate of the accumulators must be f_{os}. We summarize the observations in Table 5.3.

Table 5.3. Data rate and word length of the IIR part in figure 5.22.

	Input	Output	accumulators
Data rate	f_{os}	f_{os}/N	f_{os}
Word length	m	$M \cdot k + m$	$M \cdot k + m$

From Tables 5.2 and 5.3 it is seen that the frequency limitation of the recursive structure based on the algorithm of figure 5.22 is relaxed by using the non-recursive algorithm of figure 5.23. In the recursive structure, the IIR part has to operate at the oversampling rate and have a word length of $M \cdot k + m$. In the non-recursive structure based on the algorithm of figure 5.23, the word length is only m bits when the data rate is the oversampling rate. The word length is short when the data rate is high; and when the word length increases the sampling rate decreases. Therefore, the speed can be increased and the power dissipation can be saved by using the non-recursive algorithm of figure 5.23.

C. Comparison implementation of comb filters using a synthesis tool

Two test chips according to the two algorithms were designed in Ericsson's 3-V 0.6-μm CMOS process [68]. Both of them are programmable. We only use the high-level synthesis tool in COMPASS without resorting to any optimization. To make a fair comparison, we also consider the overhead needed to make the algorithm of figure 5.23 programmable.

In table 5.4, we compare the highest operation frequency for the two test chips. It is seen from this table that the algorithm of figure 5.23 can operate at a higher frequency. The non-recursive algorithm can increase the operation frequency by 50~75%, depending on the order k of the comb filter and decimation ratio N. The larger the filter order k and the decimation ratio N are, the more effective the non-recursive algorithm is.

Table 5.4. The maximum operation frequency. V_{dd} = 3.3 V.		
	non-recursive algorithm of figure 5.23	recursive algorithm of figure 5.22
$k = 3$, $N = 32$	120 MHz	79 MHz
$k = 3$, $N = 64$	120 MHz	77 MHz
$k = 3$, $N = 512$	120 MHz	71 MHz
$k = 5$, $N = 32$	110 MHz	72 MHz
$k = 5$, $N = 64$	110 MHz	70 MHz
$k = 5$, $N = 512$	110 MHz	63 MHz

In Table 5.5, we compare the power dissipation of the two test chips. The operation frequency is 60 MHz. It is seen from this table that the non-recursive algorithm can reduce the power dissipation by a factor of 1.4~2.3, depending on the order k of the comb filter and the decimation ratio N. The larger the filter order k and the decimation ratio N are, the more effective the non-recursive algorithm is.

Table 5.5. Power dissipation. V_{dd} = 3.3 V; operation frequency 60 MHz.		
	non-recursive algorithm of figure 5.23	recursive algorithm of figure 5.22
$k = 3$, $N = 32$	105 mW	148 mW
$k = 3$, $N = 64$	115 mW	171 mW
$k = 3$, $N = 512$	142 mW	253 mW
$k = 5$, $N = 32$	130 mW	230 mW
$k = 5$, $N = 64$	140 mW	320 mW
$k = 5$, $N = 512$	170 mW	440 mW

D. Bit-serial implementation

An viable alternative to the traditional bit-parallel arithmetic is the bit-serial arithmetic. The bit-parallel arithmetic processes W_d bits (W_d is the word length of the data) at every clock cycle, while the bit-serial arithmetic processes only one bit at every clock cycle. However, the conclusion that the bit-parallel arithmetic is about W_d times faster than the bit-serial arithmetic is not true, because the long carry propagation paths in the bit-parallel arithmetic dictate the highest clock frequency. By using the bit-serial arithmetic, we can have very compact processing elements and reduce the wiring by eliminating wide buses. In fact, the computational throughput per chip area may be higher for the bit-serial arithmetic than for the bit-parallel arithmetic, depending on the algorithm [69].

A dedicated comb filter with $k = 3$ and $N = 32$ was designed and implemented in a 5-V digital CMOS process [17, 70]. It was done by using the full-custom approach.

In Table 5.6, we summarize the measured power consumption. It is evident from this table that the non-recursive algorithm of figure 5.23 yields a low power realization.

Table 5.6. Measured power consumption of the bit-serial comb filter with $k = 3$ and $N = 32$. $V_{dd} = 5$ V.

Input data format	Clock frequency 10 MHz	Clock frequency 20 MHz
0, 0, 0, 0, ...	1.7 mW	3.45 mW
1, 1, 1, 1, ...	2.3 mW	4.75 mW
0, 1, 0, 1, ...	3.37 mW	5.43 mW
Pseudo random	~ 3 mW	~ 5 mW

5.10. OTHER AUXILIARY CIRCUITS

A. Clock voltage doubler

In Chapter 3.6 we have discussed the diode pumping technique to level-shift clock voltages in order to reduce the switch-on resistance for high-speed SI circuits. However, this technique is not effective when the supply voltage is reduced to 1.2 V since there is not so much voltage overhead. Clock voltage doublers are needed to drive the N-type switches for the 1.2-V SI circuit shown in figure 3.16. In figure 5.24, we show two clock voltage doubler circuits that can be used for this purpose.

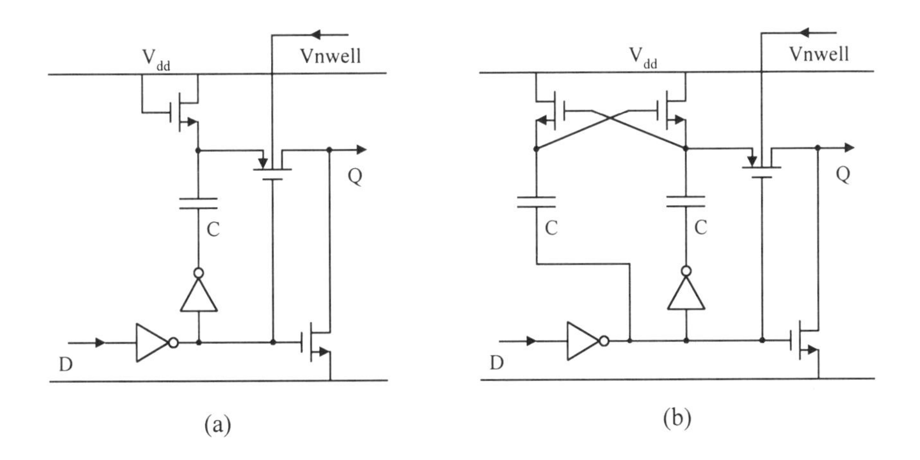

Fig. 5.24. Clock doubler circuits. Left: traditional and right: cross-coupled.

Shown in left-hand side of figure 5.24 is the conventional clock voltage doubler circuit. The drawback of the circuit is the voltage drop across the NMOS transistor. Shown in the right-hand side of figure 5.24 is the clock voltage doubler circuit using a cross-coupled transistor pair. Due to the cross coupling, the sources of the cross-coupled transistor pair are charged to the supply voltage without any voltage drop.

By applying a square wave input signal of 1.2 V at the input D, the two capacitors (formed by sandwiching poly and metal plates) are self-charged to the supply voltage 1.2 V through the cross-coupled NMOS transistors [71-73], and a square wave signal is generated at the output Q. The high voltage of the square wave signal is given by

$$V_Q = 2V_{dd} \cdot \frac{C}{C_{load} + C + C_p} \tag{5.14}$$

where V_{dd} is the supply voltage, 1.2 V, C is the capacitance connected with the cross-coupled pair, C_{load} is the load capacitance, i.e., the gate capacitance of the switch, and C_p is the parasitic capacitance at the output.

In figure 5.25 we show the simulation results of the two clock doubler circuits. In the simulation, both load and parasitic capacitances are considered.

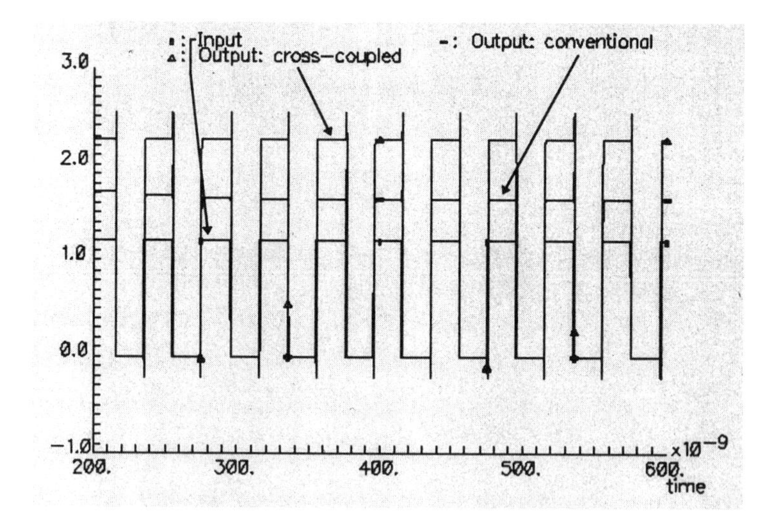

Fig. 5.25. Simulation results of the clock doubler circuits of Fig. 5.24.

It is seen in figure 5.25 that the cross-coupled circuit pumps the voltage to around 2.2 V while the conventional circuit only increases the voltage to about 1.6 V. With a clock voltage of 2.2 V, we can open the NMOS switches completely for the 1.2-V SI circuits since the DC voltages at sources of the switches transistors are low (~0.4 V).

B. High voltage generator

The bias voltage for the P-type transistor in the cross-coupled clock doubler circuit needs to be generated internally. The internal high voltage generator [71, 72] is shown in figure 5.26. It needs a square input signal.

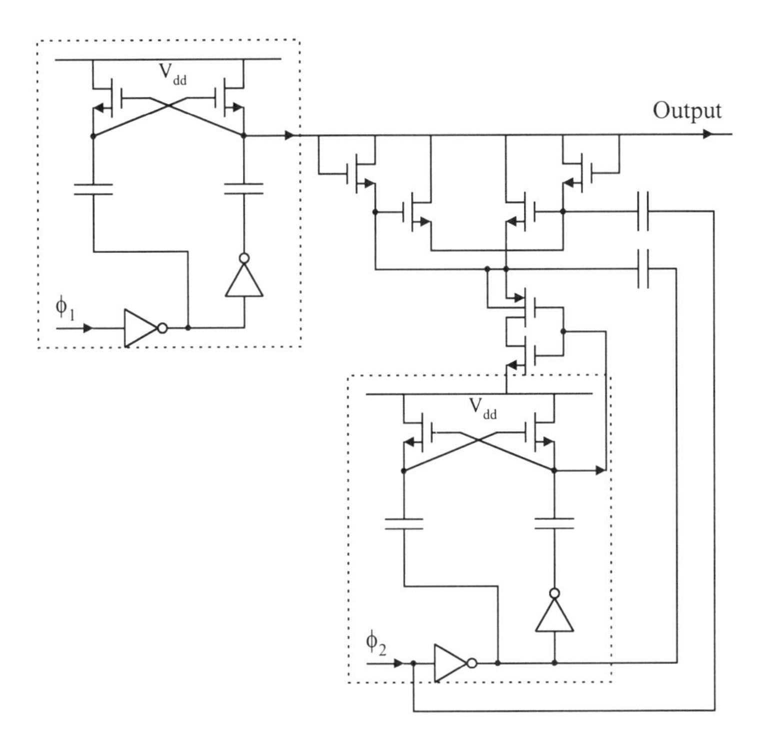

Fig. 5.26. The high voltage generator.

All the capacitors are formed by sandwiching poly and metal plates. The key feature of the high voltage generator is to use a feedback to pump the voltage across the capacitors to V_{dd} (1.2 V) in the same way as in the cross-coupled clock doubler circuit of figure 2.24. Controlled by non-overlapping clock phases (ϕ_1 and ϕ_2), this high voltage generator is capable of generating a voltage of over 2 V if the input clock phases have a 1.2-V amplitude. This voltage can be used to bias the Nwell for the P-type transistors in the clock doubler circuits of figure 2.24 to avoid the latch up. No current is needed to supply. Only a small power consumption is needed when the voltage ramps up at the beginning.

The simulated response is shown in figure 5.27. Parasitic capacitance is considered in the simulation. With a 1.2-V clock input, the output voltage is pumped to about 2.2 V.

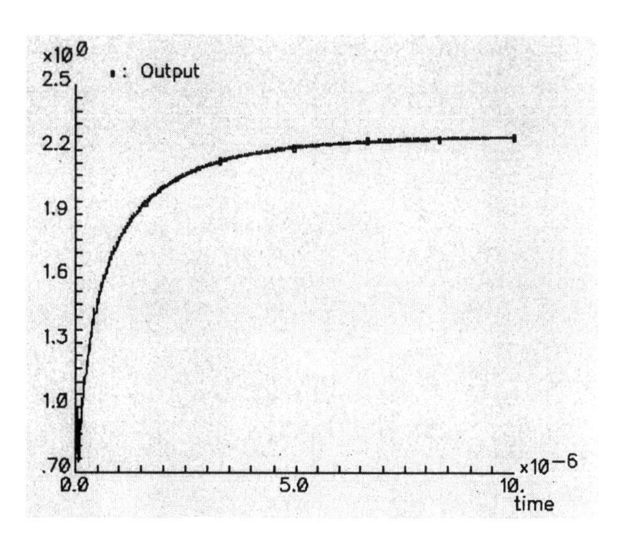

Fig. 5.27. Simulated response of the high voltage generator of Fig. 5.26.

C. Level-conversion circuit

For the oversampling A/D converter using a single 1.2-V supply voltage, the outputs of the current quantizer have a voltage swing of only 1.2 V. They cannot be directly used to control the D/A converters in that they cannot open NMOS switches completely. To interface with computers, we also need to convert the 1.2-V voltage swing of the digital outputs to a higher voltage swing, say 3 V. In figure 5.28, we show the level-conversion circuit [17].

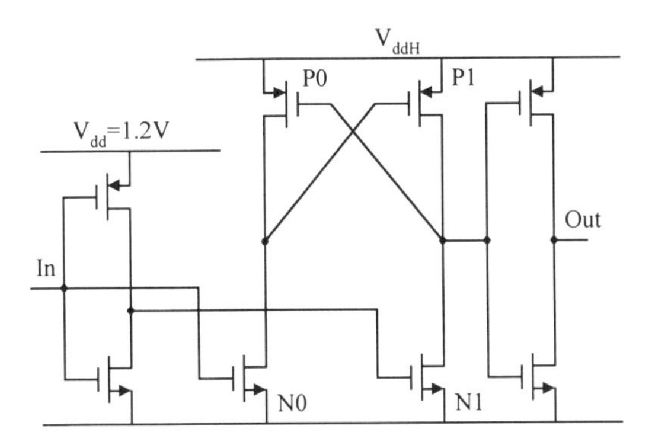

Fig. 5.28. Level-conversion circuit.

The input digital signal has a voltage swing of 1.2 V and the output digital signal has a voltage swing equal to V_{ddH}. The PMOS cross-coupled pair (P0 and P1) are connected to the high supply voltage, driven differentially (via NMOS pair N0 and N1) by low-voltage swing signals. A minimum-size inverter is used as a load. The pull-down N-devices, N0 and N1, are DC ratioed against the cross-coupled pull-up P-devices, P0 and P1, so that a low-swing input guarantees a correct output transition.

In figure 5.29, we show the simulation result of the level-conversion circuit of figure 5.28.

It is seen in figure 5.29 that we can generate a 3-V square wave output with a 1.2-V square wave input.

This level-conversion circuit consumes power only during transitions and it consumes no DC power. This level-conversion circuit only works with Nwell or twin-well CMOS processes since it requires isolated wells for the PMOS devices that are connected to the high voltage supply and the ones that are connected to the low voltage supply.

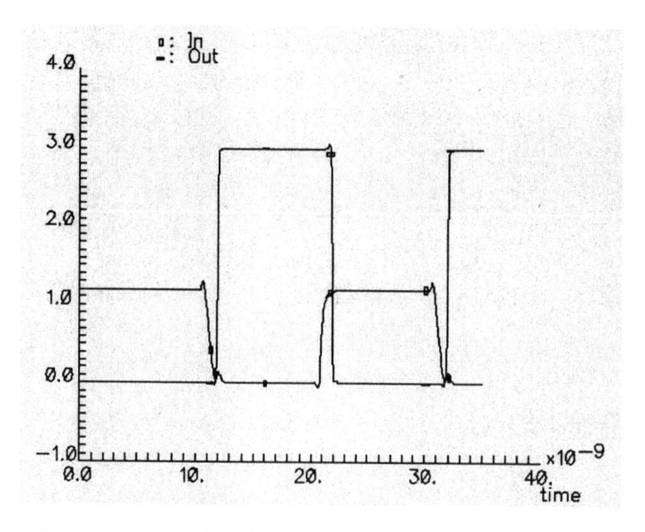

Fig. 5.29. Simulation result of the level-conversion circuit of figure 5.28.

5.11. SUMMARY

In this chapter, we have discussed different building blocks for SI oversampling A/D converters including both analog and digital circuits. We have shown SI integrators and SI differentiators and demonstrated how to use the practical SI circuits of Chapter III to construct lossless SI integrators and differentiators. We have discussed different current quantizers, from the traditional voltage comparators, to the low-input-impedance current quantizer and the resettable ultra low-voltage current quantizer. We have discussed different 1-bit D/A converters that behave as a digitally controlled current sources. To facilitate application of the SI technique, on-chip V/I converters have been discussed. We have paid special attentions to canceling off-set currents in V/I converters. We have discussed low-cost methods to realize on-chip low-pass anti-aliasing filters for oversampling A/D converters in order to realize a whole real-time signal processing system in a digital CMOS process. A combination of V/I converters and low-pass filters in a digital CMOS process has also been demonstrated. SI circuits can operate at very high speed and/or at very low supply voltage. It is challenging to design digital decimator filters that can operate at high speed with a low supply voltage and low power consumption. We have discussed the digital decimation filter algorithm and showed how to increase the operation speed and reduce power dissipation. Other auxiliary circuits such as clock voltage doublers, high voltage generators, and level-conversion circuits have also been covered in this chapter.

Chapter VI: Practical Aspects of SI Circuits and Systems

6.1. Introduction

In this chapter, we will discuss practical aspects of SI circuits and systems. To predict the performance or even verify the functionality of an SI circuit, a proper simulation setup is needed. Due to the high output impedance of SI circuits, the outputs of SI circuits should not be left open-circuited. A proper clocking and loading are the ways to avoid open-circuiting the outputs of SI circuits. To facilitate the convergence at the simulation start-up and to avoid large start-up impulse currents within SI circuits, the resetting can be used. To implement SI circuits, layout issues need to be addressed. When we implement both analog and digital circuits on the same chip (as in SI oversampling A/D converters), the noise coupling via the parasitic capacitance, power supplies, and substrate is an important issue that will also be discussed in this chapter.

6.2. Simulation Setup

To start a design, some people tend to use some ideal components (e.g., ideal current sources) to check the functionality of a circuit. This is not good for SI circuits, especially for the second-generation SI circuits. Shown in figure 6.1 is the basic second-generation SI memory cell with an input current source.

During clock phase ϕ_1 when switches S0 and S1 are closed, the input current I_i is the input to the memory transistor M0. During clock phase ϕ_2 when switch S2 is closed, the output current I_o is generated. However during clock phase ϕ_2, the input current source I_i is left open-circuited and the potential can ramp up to a very high value, much higher than the supply voltage. Due to the numerical convergence in the simulator (e.g., SPICE), this very large voltage can change the drain potential of memory transistor M0 though the gate voltage for the switch S1 is low. Therefore, the current source I_i has to be terminated during clock phase ϕ_2.

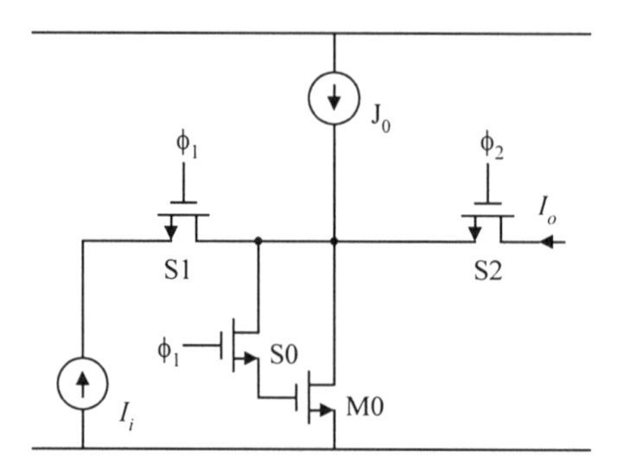

Fig. 6.1. Basic second-generation SI memory cell with an input current source.

In figure 6.2, we show the SI memory cell with an input termination. The termination is provided for the input current source I_i on clock phase ϕ_2. The termination is a copy of the SI memory cell with the voltage-sampling switch S3 always closed. Therefore, the current source I_i always has a low-impedance load on both clock phases.

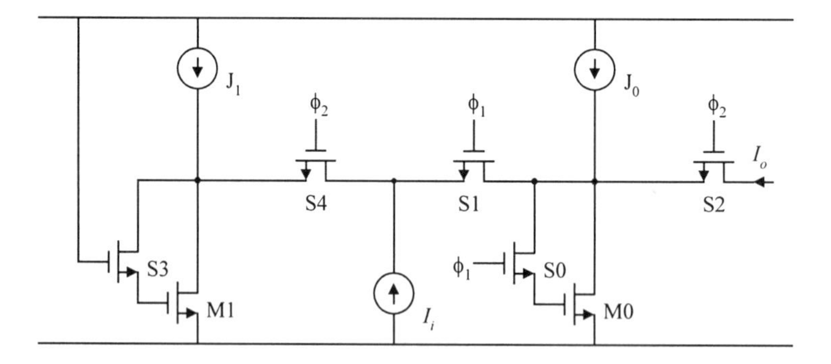

Fig. 6.2. Basic second-generation SI memory cell with an input termination.

During the non-overlapping interval, however, the potential at the input current source can still change dramatically. It is better to reduce the output impedance of the input current source to a value that is a little bit higher than the output impedance of the memory cell. By doing so, the change of the potential at the input current source can be reduced and the limited output impedance will only introduce a negligible error compared to the error introduced in the memory cell due to its own finite input-output conductance ratio. The clocking scheme to be discussed in Chapter 6.3 should be used to further reduce the potential change at the output node of the input current source.

For the first-generation SI circuits (see figure 1.1 (a)), there is always a current mirror at the input, we do not have the problem outlined above.

To simulate SI circuits, the simulator must have the charge conservation feature. The simulation algorithm should be 'gear' instead of 'trapezoidal' to avoid ringing due to the numerical error. When the accuracy is needed, the simulation option 'accurate' in HSPICE can be chosen. It gives more accurate simulation results, but it takes much more time.

6.3. Clocking

Non-overlapping clocks are widely used for SI circuits [1] as well as for SC circuits. The use of non-overlapping clock in the second-generation SI circuits introduces circuit errors due to transient spikes [31]. A modified clocking strategy can reduce the transient spikes, thus minimizing the circuit errors [17, 21].

There are two kinds of switches in SI circuits. One is the voltage-sampling switch and the other is the current-steering switch. The voltage-sample switch samples a voltage on the gate of a memory transistor, while the current-steering switch directs the current flow. A non-overlapping clock should be used for voltage-sampling switches just in the same way as in SC circuits. If a non-overlapping clock is also used for current-steering switches, there is a time interval when the current source has no low-impedance load. This temporary open circuiting of the current source creates a transient spike that can couple onto the gate capacitor via the drain-gate parasitic capacitance of the memory transistor. Even though the drain voltage will settle into the same value eventually, the spike has a different fall and rise rate and the parasitic capacitance is signal dependent and highly nonlinear. Therefore circuit errors result [31]. We can use the clocking scheme illustrated in figure 6.3 to minimize the errors due to the transient spikes.

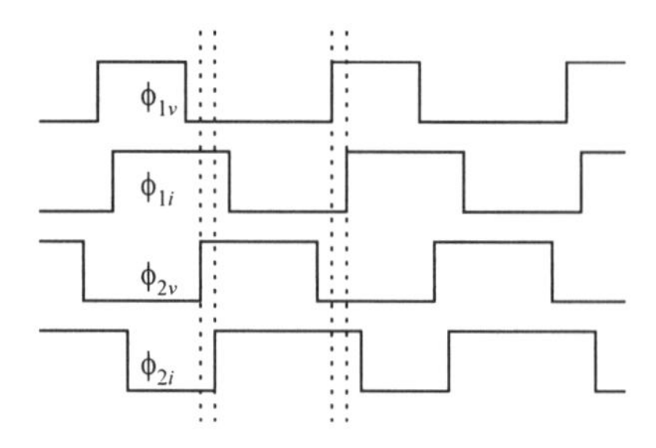

Fig. 6.3. Clocking strategy for SI circuits.

The non-overlapping clock phases ϕ_{1v} and ϕ_{2v} are for voltage-sampling switches and the overlapping clock phases ϕ_{1i} and ϕ_{2i} are for current-steering switches. The overlapping clock phases ϕ_{1i} and ϕ_{2i} (for current-steering switches) lag the non-overlapping clock phases ϕ_{1v} and ϕ_{2v} (for voltage-sampling switches), respectively. This means that current-steering switches always direct currents to low-impedance nodes formed by closing corresponding voltage-sampling switches a little bit earlier. This clocking strategy guarantees that during the circuit operation, no current will be directed to a high-impedance node. The transient spikes decrease significantly and the circuit errors due to the transient spikes are thus minimized.

For S^{2}I circuits, we can still use this clocking strategy. In this case, we only need split each non-overlapping phase into two non-overlapping phases to control the two-step memorization. The overlapping clock phases can still be used to reduce the transient glitches.

The non-overlapping and overlapping clock generators are shown in figures 5.10 and 5.11 in Chapter 5.6, respectively.

6.4. LOADING

For the second-generation SI circuits, the output of the current mirror realizing a scaling coefficient is usually only needed during one clock phase. If we leave the output open circuited on the other clock phase, the output potential will change drastically to the supply voltage or ground. This potential change will introduce errors via the drain-gate parasitic capacitance as in the case where non-overlapping clock phases are used for current-steering switches. Therefore, during the clock phase when the output is not valid (not connected with the following circuits), the output should be connected to a low-impedance node to avoid the drastic potential change at the output. It is of the same consideration as for the simulation setup discussed in Chapter 6.2. The low-impedance node can be created by a memory cell on its input phase. This is shown in figure 6.4.

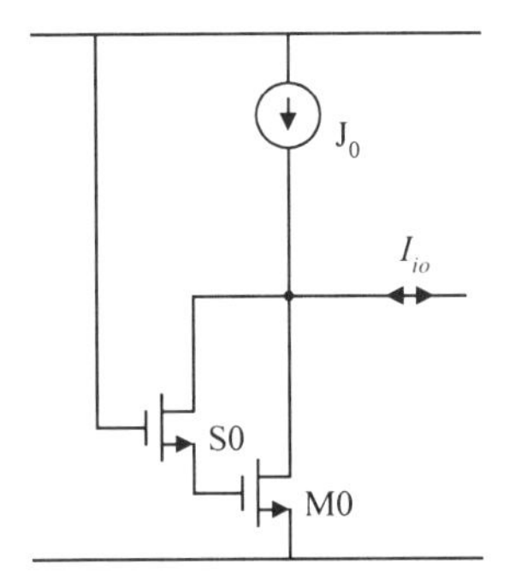

Fig. 6.4. Load for terminating SI circuit outputs on the clock phase when the output currents are not needed by the following circuits.

Shown in figure 6.4 is an SI memory cell with the voltage-sampling switch S0 always closed. Therefore, a low-impedance node is created. The memory transistor size and the bias current of the load should be the same (or scaled by the same factor) as the memory transistor size and the bias current of the main SI circuits, respectively. By doing so, the input potential of the load is the same as the potential of the main SI circuits with a zero input current. The bias current (and the memory transistor size) of the load can be designed so that even with the largest input current, the input potential of the load is still comparable to the corresponding potential in the main SI circuits.

For fully differential SI circuits, both the fully differential outputs can be directed to the same load. Since we direct the fully differential outputs to the load, the actual input current to the load is small and therefore only a small bias current for the load is needed.

We can ether use one load for all the SI circuit blocks or use different loads for different SI circuit blocks. The choice should be made with consideration of the layout and floor planning. For practical circuits, the basic memory cell circuit used in the load in figure 6.4 should be replaced by the practical SI memory cell circuits used in the SI system.

6.5. Resetting

When we simulate a large SI circuit, it is not unusual to have a convergence problem due to the fact that some voltage-sampling switches are open and the gates of the associated memory transistors are floating. To avoid this problem, we usually set initial voltages to make sure that all the voltage-sampling switches are closed and that there are no floating gates.

If we leave the gate of a transistor open-circuited at the start-up, the potential can be anything. This may make the drain current very large. Theoretically, this just behaves as an impulse input and it will vanish after a while. However, a very large start-up current may destroy all the proper operation and the feedback current in an oversampling delta-sigma modulator may not be effectively large enough to make the circuit eventually settle into the proper operation. For real telecommunication applications, we usually require power up and power down features. The excess start-up time at the power up due to the start-up impulse current may not be acceptable.

To avoid the convergence problem in the simulation and the problem in reality outlined above, we can use an extra signal to reset all the SI circuits. The reset signal can also be generated by a power-up signal. We provide a resetting signal which can close all the voltage-sampling switches and open all the current-steering switches. The outputs of the D/A converters in the delta-sigma modulator are short-circuited at the start-up as well.

Another advantage of the resetting is the possibility of getting rid of some non-functioning circuits (due to the yield of the process) without resorting to the expensive performance testing. When the reset signal is active, the supply current is well defined. By measuring the supply current we can identify non-functioning circuits on which we do not need to do a further performance testing.

The resetting signal can directly control some start-up switches that set the gates of memory transistors to proper potentials. The resetting signal can also be combined with the clock generators of figures 5.10 and 5.11. An example of a clock generator for SI circuits [74] is shown in figure 6.5.

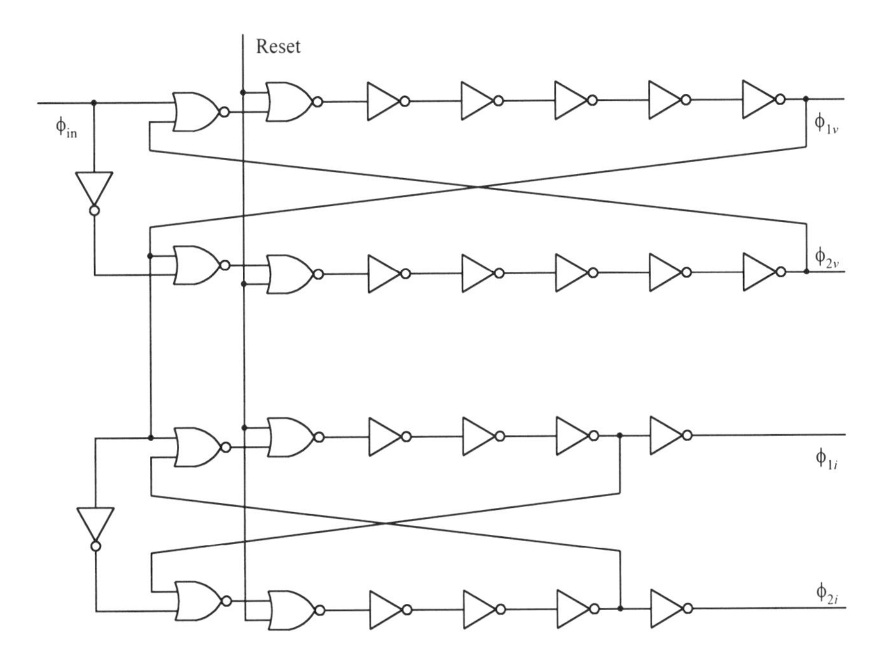

Fig. 6.5. An example of a complete clock generator for SI circuits.

The two clock phases ϕ_{1v} and ϕ_{2v} for voltage-sampling switches are non-overlapping and the two clock phases ϕ_{1i} and ϕ_{2i} for current-steering switches are overlapping. The clock phase ϕ_{1v} is used to generate the clock phases for current-steering switches so that the timing relationship between all the clock phases is as illustrated in figure 6.3. When the reset signal is active (high), the clock phases ϕ_{1v} and ϕ_{2v} for voltage-sampling switches are high, and the clock phases ϕ_{1i} and ϕ_{2i} for current-steering switches are low. The overlapping and non-overlapping intervals can be controlled by the number of inverters (i.e., delay) in the chains. The timing relationship between the non-overlapping clock phases and overlapping clock phases can be controlled by the input to the overlapping clock generator in respective of the output of the non-overlapping clock generator.

6.6. Basic Analog Layout

A. Layout of analog components

In a digital CMOS process, we can create resistors, capacitors, and transistors. Resistors can be used in on-chip V/I converters (see Chapter 5.7) and/or low-pass filters (see Chapter 5.8). In figure 6.6, we show a resistor created by the poly layer.

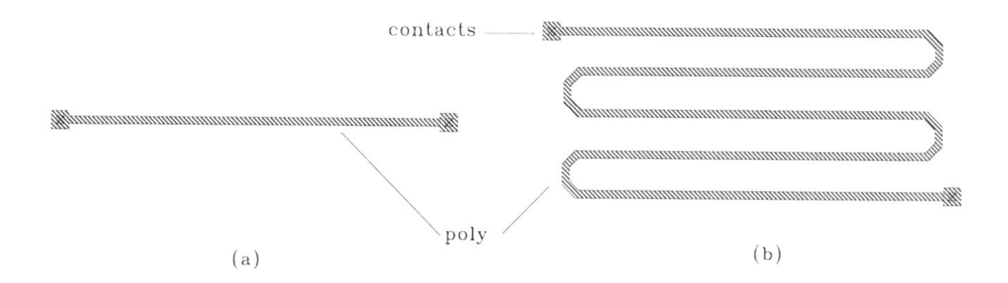

Fig. 6.6. Layout of a poly resistor of (a) a single strip and (b) multiple strips.

The resistance R is given by

$$R = 2R_c + \frac{W}{L} \cdot R_q \tag{6.1}$$

where R_c is the localized resistance of the contact between the ends of the resistor and the metal connection, W is the width of the poly strip, L is the length of the poly strip, and R_q is the sheet resistance of the poly strip.

Since the sheet resistance of the poly strip in a modern digital CMOS process is very low (1~2 ohms/square) and the variation is very large (100~200%), poly resistors are not often used. If the poly resistors are used, care should be taken to account for the parasitic resistance of the metal wires in calculating the resistance due to the small sheet resistance value of the silicidated poly layer. When a large resistance, i.e., a large aspect ratio (W/L) is needed, it is a good practice to use the multi-strip layout style shown in figure 6.6 (b).

Resistors with high sheet resistance values can be created in a digital CMOS process by using the well layer. The well layer usually has a sheet resistance around 1 kohms/square. But this kind of resistors are nonlinear and noisy and have large parasitic capacitance to the substrate. The nonlinearity is due to the signal-dependent PN junction depth which changes the resistance. The large parasitic capacitance is also due to the PN junction between the well and the substrate. This large parasitic capacitance to the substrate makes the well quite noisy since all the disturbance and noise can be coupled directly onto the well resistor. We can reduce the noise coupling via the substrate by surrounding the well resistors with a biased well ring and a biased substrate ring. This kind of well resistors can be used in the applications where the performance is not so demanding (e.g., in Chapter 5.8). In figure 6.7, we show an Nwell resistor in a P-substrate CMOS process.

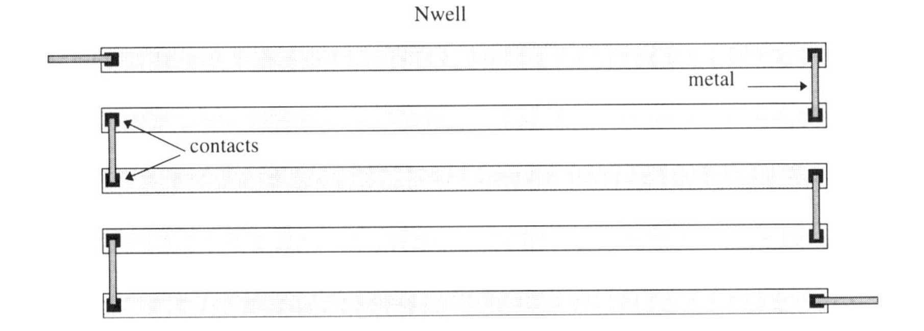

Fig. 6.7. Layout of an Nwell resistor in a P-substrate process.

To calculate the resistance, we can still apply equation (6.1) and the resistance of the contacts (~1 ohm) and the metal connections can be neglected.

When linear or floating capacitors are needed as in clock voltage doublers (see Chapter 5.10), they can be created by sandwiching the conducting layers (poly and all the metal layers). We show a linear capacitor created in a single-poly four-metal CMOS process. The bottom plate consists of the poly, metal 2 and metal 4 layers (metal 4 is the highest level metal layer) and the upper plate consists of metal 1 and metal 3. Notice that metal 4 is usually considerably thicker than other metal layers.

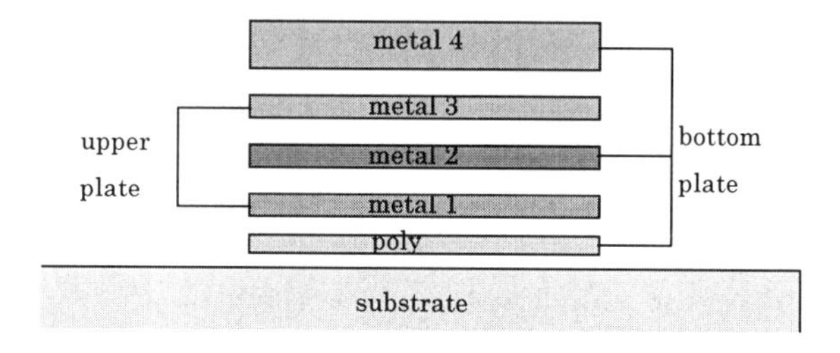

Fig. 6.8. Layout of a linear capacitor in a single-poly four-metal digital CMOS process.

If we neglect the fringing capacitance, the total capacitance C can be approximated by

$$C = 4 \cdot WL \cdot C_q \tag{6.2}$$

where (WL) is the area of the capacitor and C_q is the sheet capacitance of the parasitic capacitor between adjacent layers (assumed to be the same). Since the value of C_q is much smaller than the sheet gate capacitance, this kind of capacitors are very area consuming.

Large parasitic capacitance of the bottom plate is also present, dominated by the poly to substrate capacitance. It is given by

$$C_{bp} = WL \cdot C_{poly} \tag{6.3}$$

where C_{bp} is the parasitic capacitance of the bottom plate of the linear capacitor of figure 6.8 and C_{poly} is the sheet capacitance of the poly-to-substrate capacitors. Notice that C_{poly} is usually considerable larger than C_q, the parasitic capacitance at the bottom plate may be as much as 30~40% of the total capacitance.

Also noted is that the matching and linearity of this kind of capacitors are not as good as the matching and linearity of poly-poly capacitors due to the use of the thick oxide as the dielectric layers and the roughness of the metal surface.

The most commonly used devices in a digital CMOS process are the MOS transistors. For SI circuits, the drain-gate parasitic capacitance introduces errors. It is of importance to minimize the drain area. This can be achieved by splitting a single transistor into a number of transistors in parallel connection as shown in figure 6.9.

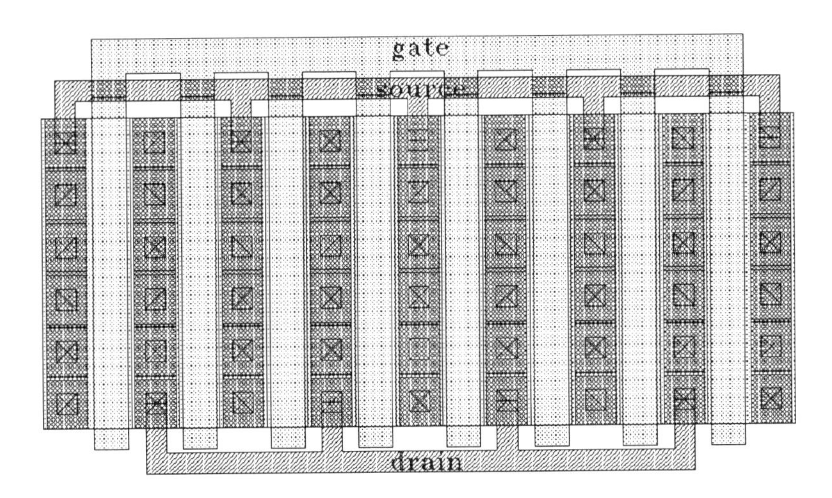

Fig. 6.9. Layout of a transistor to minimize the drain area.

Eight transistors are connected in parallel in figure 6.9. All the gates are connected by the poly layer and all the source and drains are connected by a metal layer, respectively. To reduce the source and drain resistance, as many contacts as possible are used.

Due to the parallel connection, the drain area is reduced by a factor of two and the parasitic gate resistance is reduced approximately by a factor of eight. The layout style illustrated in figure 6.9 is a good practice for high speed, high accuracy applications due to the reduced parasitic gate resistance and parasitic drain capacitance.

It is also of importance try not to let the gate and drain cross over each other as shown in figure 6.9 to minimize the drain-gate capacitance.

B. Matching considerations

Matching of analog components is of great importance, though in oversampling delta-sigma modulators the requirement on matching is relaxed. To increase the matching, it is usually beneficial to split the components into many unit cells (preferably with the same size) and try to matching each and every corresponding cell. There are two ways to increase the matching. One is the interdigitized layout style and the other is the common-centroid layout style.

By using the interdigitized layout style, the process variation at one direction can be minimized. Shown in figure 6.10 is the interdigitized layout style applied to a poly resistor pair.

The process variation on the vertical direction has the same effect on the resistor pair and therefore, its influence on the matching is minimized. To reduce the boundary effect [75], two poly strips are placed beside the poly resistor pair. Since the ploy resistors have a very small sheet resistance, it is important to match the metal wires that connect the individual resistor strips.

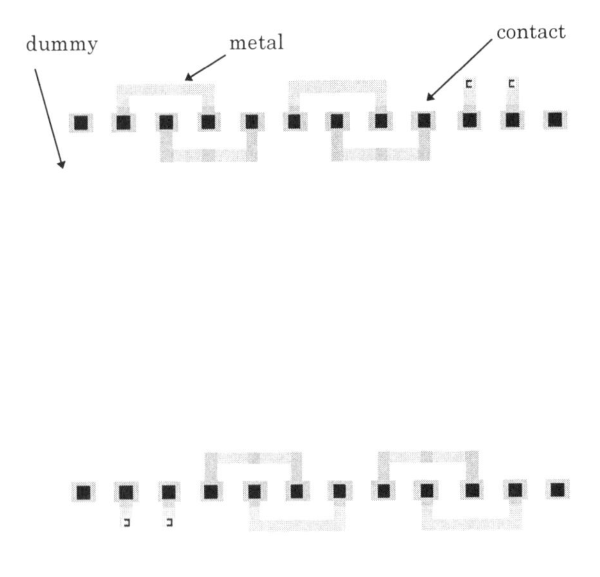

Fig. 6.10. Interdigitized layout style applied to a poly resistor pair.

To reduce the influence of the process variation on both the horizontal and vertical directions, we can use the common-centroid layout style. Shown in figure 6.11 is the common centroid layout style applied to a linear capacitor pair created by sandwiching poly and metal layers as shown in figure 6.8.

Dummy strips around the capacitor arrays are used to reduce the boundary effect. A biased Nwell (the substrate is of P type) is placed beneath the capacitor array and a bias substrate ring is placed surrounding the capacitor array to reduce the noise coupling via the substrate. Due to the small sheet capacitance, it is important to try to match the metal wires that connect the individual cells. Otherwise, the parasitic capacitance associated with the metal leads will destroy the matching. Due to the use of the common centroid layout style, the process variations on both the horizontal and vertical directions have the same effect on the capacitor pair, thereby, their effect on the matching is minimized.

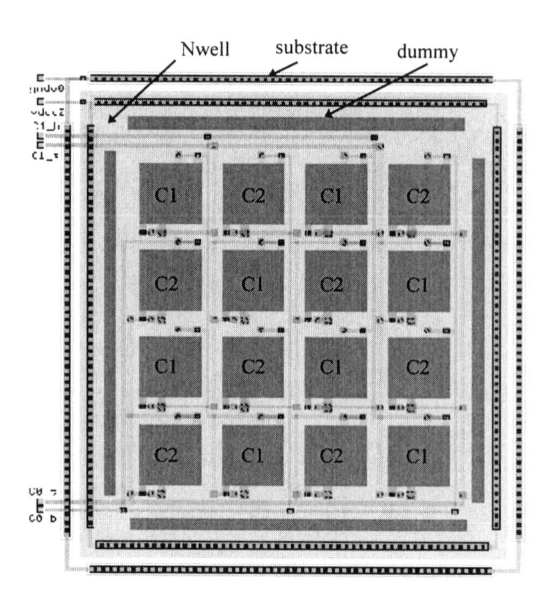

Fig. 6.11. Common centroid layout style applied to a linear capacitor pair.

The most commonly used devices in a digital CMOS process are the MOS transistors. We have already discussed in Chapter 2.2 that the parasitic resistance has a very important influence on the matching. After having considered the mismatch due to the parasitic resistance, we have to deal with the variation in the physical parameters.

In order to match transistors, we should first guarantee that the transistors have the same orientation and the same current flow direction. The two commonly used matching techniques interdigitized and common-centroid can also be applied to the transistor matching. In figure 6.12, we show the layout of a transistor pair. Each transistor contains four unit transistors in parallel.

Shown in figure 6.12 (a) is the normal way to layout the transistor pair without any consideration of matching. By using the interdigitized layout style illustrated in figure 6.12 (b), matching can be improved compared with the normal layout style illustrated in figure 6.12 (a) in that the segments of transistors needed to be matched are closer. In the common-centroid layout shown in figure 6.12 (c), the centroid of both composite devices lies at the center of the structure. Therefore, the process variations on both horizontal and vertical directions have the same effect on the transistor pair. The common-centroid layout style usually yield a better matching than the interdigitized layout style especially when it is applied to passive components. However, for transistors having three individual terminals, the common-centroid layout style may not yield the best matching due to the complexity, relative irregularity, and increased distance between the devices. If the transistors to be matched share some common-terminals (e.g., the same bias voltage, or the same source potential as in a fully differential pair), the complexity of the common-centroid layout is reduced significantly and the common-centroid layout style can have a better matching than the interdigitized layout style.

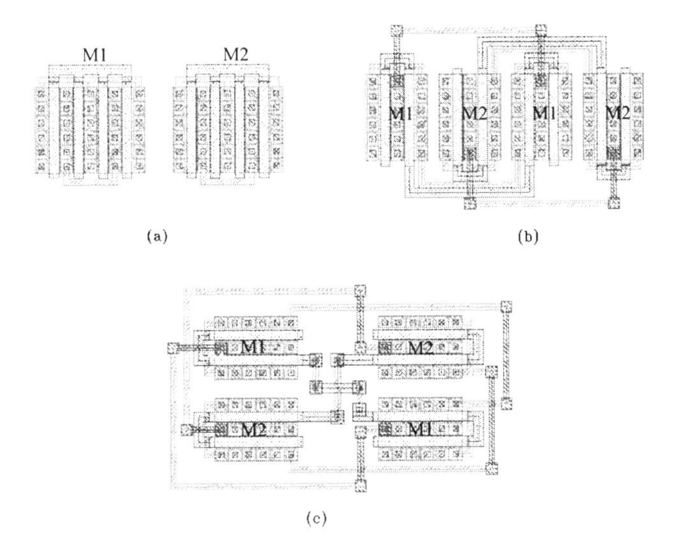

Fig. 6.12. Layout of a transistor pair. (a) normal, (b) interdigitized, and (c) common-centroid.

In SI circuits, we do not only need to match a transistor pair, but also need to match the whole memory cell architecture and their associated current sources. This makes the common-centroid layout style difficult to apply. It is difficult to match all the parasitic associated with the wiring. If we use the common-centroid layout style, the layout can not be made regular in the sense both directions have to be expanded when we change the scaling factor. Therefore, it is better to use the interdigitized layout style which usually results in a very regular layout. When we change the scaling factor, the size on only one direction needs to be changed and therefore, we can maintain the constant pitch of power supplies. This does not only makes the layout regular but also simplifies the floor planning and placement.

In figure 6.13, we show an example of using the interdigitized layout style applied to a fully differential SI memory cell circuit consisting of 8 unit cells.

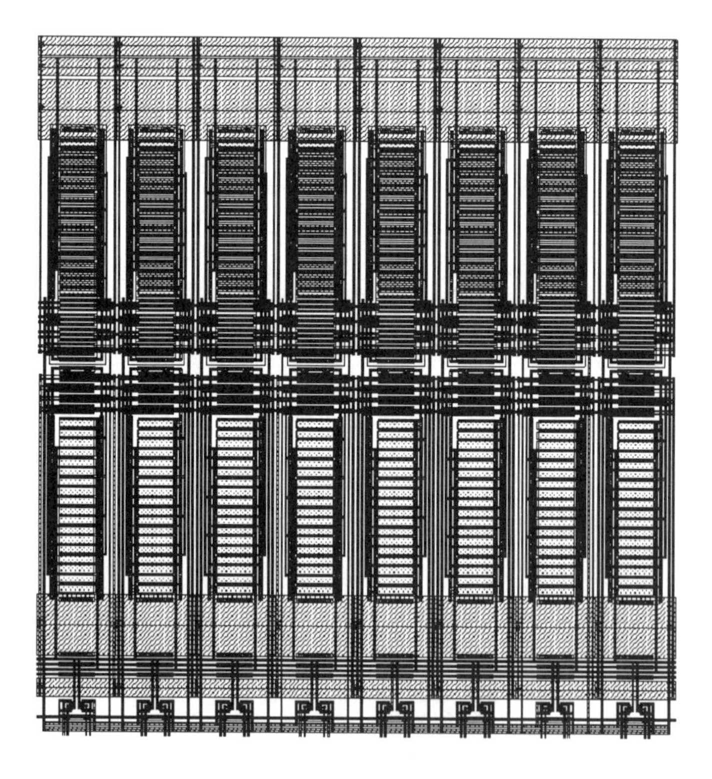

Fig. 6.13. An example of the interdigitized layout style applied to a fully differential SI memory circuit.

In figure 6.13, each cell is a fully differential SI memory cell consisting a cascoded NMOS memory transistor pair and a cascoded P-type current source pair. Shown in the bottom are the switch transistors and clock busses. When different scaling factors are needed, we only need change the horizontal direction by adding or deleting unit cells. The interdigitized layout style is especially effective when we use the current scaling principle and different scaling factors are needed. It results in a very regular layout

Sometimes, we can even combine the common-centroid and interdigitized layout styles. We can use the common-centroid layout style locally for the best matching and use the interdigitized layout style globally for regularity. Such an example is shown in figure 6.14.

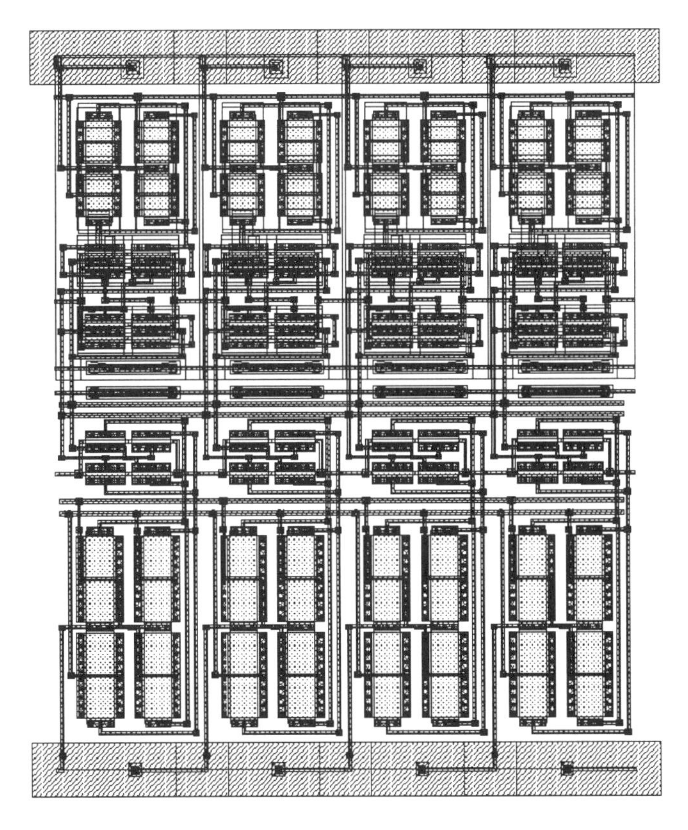

Fig. 6.14. An example of combining the common-centroid and interdigitized layout styles in a fully differential SI memory circuit.

The common-centroid block is a unit cell in which all the transistors including memory transistors, current source transistors, and cascode transistors are laid out with the common-centroid layout style to increase the matching. All the unit cells (four in figure 6.14) are interdigitized to maintain the matching and increase the regularity. When different scaling factors are needed, we only need to change the number of the unit cells in parallel.

The boundary dependent etching of polysilicon gates [75] reduces the matching at the boundary as it has the same effect on the passive component matching. Due to the boundary effect, dummies should be used at the boundary of the components that need to be matched. Make sure that all these dummy transistors are cut-off. It is also a good idea to use these dummy transistors as local de-coupling capacitors.

6.7. Mixed Analog-Digital Layout

When we integrate both analog and digital circuits on the same chip, the noise coupling via the wire capacitance, power supply lines, and substrate may limit the performance. In the following discussion, we assume that the substrate is lightly doped (resistivity ~10 ohm-cm). We will also touch upon the noise coupling for a heavily-doped substrate (resistivity ~5 mohm-cm).

A. Protection of sensitive analog signals

Integrating both analog and digital circuits on the same chip, we need to protect the sensitive analog signals from the noisy environment. In oversampling A/D converters, the most sensitive part is at the input. A general practice to shield sensitive analog signals is shown in figure 6.15.

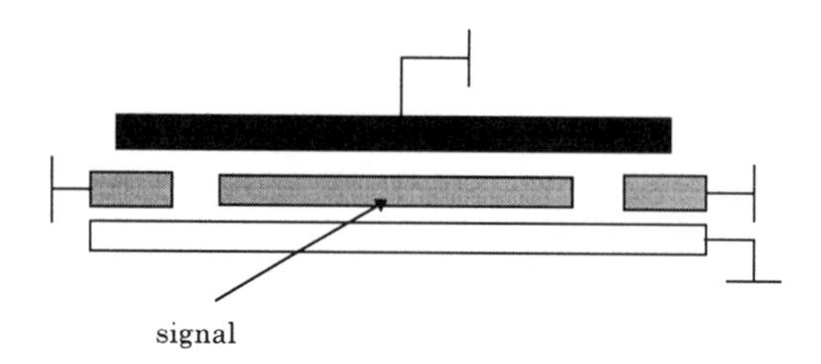

Fig. 6.15. Shielding of the sensitive analog signal.

The conducting layer carrying the sensitive analog signal is surrounded by grounded conducting layers. Noise, therefore, cannot couple onto the sensitive analog signal. However, it is very important to guarantee that the ground supply is clean and free from noise. Therefore, the ground is preferably supplies by an extra pin. If not possible, the ground should only be connected with the analog ground at the circuit boundary, i.e., close to the pads.

In SI circuits, the signals are currents and they are always directed to low impedance nodes. Compared with the charge being temporarily stored at the gates of memory transistors, the currents are much less sensitive to noise. Therefore, special attention must be paid to avoid the coupling of noise onto the gate of a memory transistor when the SI memory cell is in its hold mode.

First of all, under no circumstances should the clock signals cross the gates of the memory transistors. Shielding of the gates of memory transistors is recommended. Due to the switching in SI circuits, the currents are noisy as well and they should not cross the gates of memory transistors as well. In figure 6.16, we show a conceptual protection of the gates of memory transistors. The memory transistor is surrounded by a grounded guard ring.

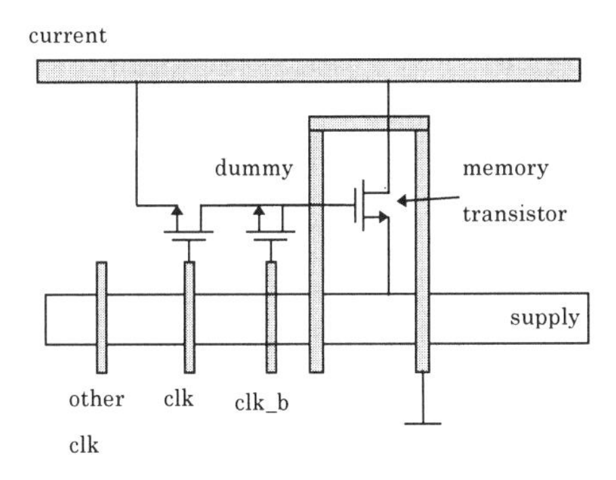

Fig. 6.16. Shielding of sensitive analog parts in SI circuits.

When a current mirror follows the memory cell, we can put both the memory transistor and current mirror transistor in the same protection. Sometimes to make the layout easier, we can even put the associated cascode transistors in the same protection. To avoid the coupling via the substrate, a substrate contact ring should surround the memory cell transistor. Assume the substrate is of P-type, the substrate contact ring for NMOS memory transistors should be biased at a clean V_{ss} supply (the lowest supply voltage). If PMOS transistors are used as memory transistors, an Nwell ring biased at a clean V_{dd} supply (the highest supply voltage) should surround them. To further reduce the noise coupling through the substrate, a substrate contact ring biased at a clean V_{ss} supply can be used to surround the N-well protection ring.

B. Clock distribution

In a mixed analog-digital chip, clocking is troublesome. We usually need to distribute clock signals into analog parts. Many clock signals switching simultaneously generate a lot of switching noise if clock buses are long. Therefore, it is wise to reduce the number of clock signals that need to be distributed. For example the inverse of the clock signals can be generated locally. Sometimes, when the chip is really big, even the whole non-overlapping and overlapping clock generators can be localized and therefore we only need to send one clock signal to different parts. In figure 6.17, we show a clock distribution network we usually use.

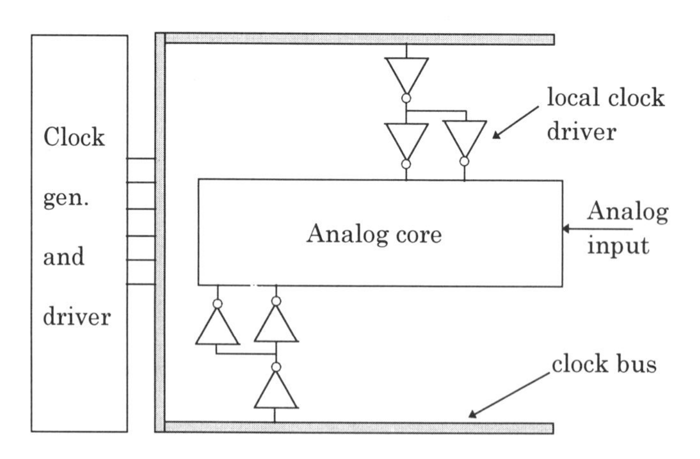

Fig. 6.17. Clock distribution network.

Clock signals are generated in the clock generator and buffered before driving the long clock buses. Local clock buffers are used to reshape the clock signals and to generate the inverse of clock signals if needed. The analog core is protected by both substrate and well contact rings. To reduce the switching noise to the substrate, the clock buses should utilize higher level metal layers and use narrow metal layers to reduce the parasitic capacitance. However, narrow metal layers have large parasitic resistance which may have a significant influence on the delay. The width of the metal layer should be several times larger than the minimum width. It is sometimes helpful to distribute the clock signals on two metal layers in order to reduce the parasitic resistance and not increase the parasitic capacitance. However, due to the fringing effect, the parasitic capacitance is likely to increase. The distance between the buses should be larger than the minimum distance required in the given process technology in order to increase the yield and decrease the coupling between the clock signals.

It is a general practice to clock a large digital VLSI chip using a tree-alike clock distribution network. If the analog part in a mixed analog-digital part is very big, we can also use a tree-alike clock distribution network to distribute one clock signal and use clock generators to generate different clock signals locally. The distribution network shown in figure 6.17 has been verified for an analog core as large as 2 mm x 2 mm by considering parasitic resistance and capacitance and worst case parameter spread. All the locally generated clock signals and their timing relationships are guaranteed. However, if the clock frequency is in the several hundred MHz range, the parasitic self and mutual inductance should be considered and the cross coupling between clock lines may destroy the timing relationship.

C. Switches

For current-steering switches, the major concern is the signal-dependent voltage drop across the switches. It is therefore important to use large transistors. Whether NMOS (or PMOS) or CMOS switches should be used is dependent on the potentials. If the potential is in the middle of the supply voltage, CMOS switches should be used. If the potential is close to the positive supply, PMOS switches should be used. If the potential is close to the negative supply, NMOS switches should be used. Since current-steering switches only connect low-impedance nodes, the noise coupling from the clock signals does not have a strong effect.

More critical are the voltage-sampling switches. Since the clock feedthrough error in CMOS switches is more dependent on the clock signals, i.e., the transition time, the timing relationship, etc. It is usually preferred to use NMOS or PMOS switches. This is especially true for SI circuits where the potential change at the gate is limited compared to SC circuits which the potential change across a switched capacitor is usually between the supply rails. By properly designing the DC voltage at the gate of the memory transistor, an NMOS or PMOS switch suffices without resorting to a CMOS switch to reduce the switch-on resistance to increase the speed. If a clock signal with a very short transition time can be generated locally, it is effective to reduce the clock feedthrough by using a dummy switch which has the half size of the main switch. Since the voltage-sampling switches connect the gates of memory transistors, it is of importance to minimize the noise coupling from the clock line to the gate capacitors of the memory transistors.

Shown in figure 6.18 is the layout of a voltage-sampling switch with and without dummy switches.

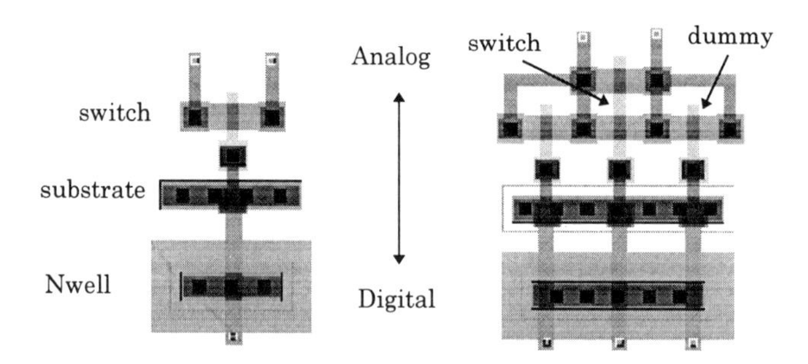

Fig. 6. 18. Layout of an NMOS switch without and with dummy switches.

It is very important not to let the digital signals cross the analog signals. Digital signals are fed from one direction and analog signals are fed from the other direction. The digital signals are carried on a high-level metal layer to reduce the switching noise to the substrate. (Notice that in certain CMOS processes, vias are allowed to stack upon contacts.) To further reduce the noise coupling, protection Nwell and substrate strips biased at clean (or analog) supplies can be used as shown in figure 6.18. To increase the matching between the switch transistor and the dummy transistor, the switch transistor is split into two transistors in parallel. It is also of benefit to use dummy transistors on both sides of the switch transistor as shown in figure 6.18.

D. Power supplies

The switching noise from digital circuits can directly coupled onto analog parts if common supplies are used. It is therefore most important to have separate analog and digital supplies in a mixed analog-digital chip.

If possible, digital output buffers should have dedicated supply lines. They are likely the most noisy supplies due to the large capacitive load. For the local clock buffers used in the clock distribution network shown in figure 6.17, the supplies for the last inverters can be connected with analog supplies. However, it might increase the difficulty in the layout.

It is also of significance to separate "noisy" analog supplies. If possible, separate supplies should be used for biasing guard or protection rings, for analog bias circuits, and for major analog blocks. Otherwise, they should be connected at the chip boundary, i.e., at the pads.

The minimum width of power supply lines is ultimately governed by the current density that the wire can sustain. The parasitic resistance usually imposes a limit far stringent than the current density requirement. The width of power supply lines should be wide enough that the voltage drop across the whole power supply line is negligible. (With reduced supply voltages, analog circuits are sensitive to the voltage drop across the supply lines.) The parasitic resistance can be estimated by using equation (6.1). With a known supply current, we can estimate the voltage drop across the whole power supply. Whether the voltage drop is acceptable is dependent on the analog circuit. If the voltage drop is not acceptable, we have to increase the wire width of the power supply line. We can also use several conducting layers as the power supply lines at the same time and feed the supplies from both ends to reduce the voltage drop in the supply lines due to the parasitic resistance.

If a large current is needed by a supply line, several pads should be used.

E. Bias circuits

Bias circuits are also very important to the performance of analog circuits. A noisy bias circuit can increase the total noise level in the main analog circuits significantly. During switching, due to the parasitic capacitance, glitches are present at the bias points. The fluctuation in the bias voltages and currents introduces distortion. The fluctuation in the bias needs to be reduced. This can be accomplished by reducing the impedance and/or adding de-coupling capacitance.

If the analog circuit is large, we need to distribute the bias signals to all the building blocks. It is better to use localized bias circuits and control them by a bias current. The voltage biasing is more sensitive to the voltage drop in the bias line due to the parasitic resistance of the wire. It is also a good practice to use separate supply lines for the bias circuit blocks and for the main circuit blocks since the supply lines for the major analog blocks are relatively noisy due to the switching in SI circuits. The supply lines can be connected together at the pads.

In figure 6.19, we show the bias arrangement for high performance analog circuits, illustrating the principle.

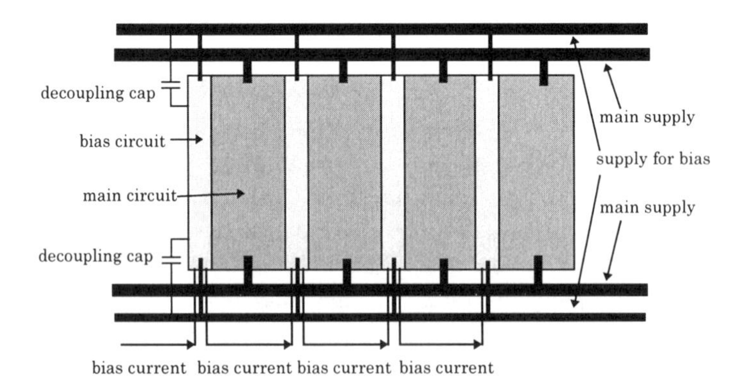

Fig. 6.19. Bias technique for high performance analog circuits.

F. Pin assignment

Due to the large capacitance and bond wire inductance, the noise coupling due to pads and pins may be dominating if not properly addressed. The pins carrying sensitive analog signals and most noisy digital signals should be most apart. The pins connecting most sensitive analog signals can even be surrounded by one ground pin on each side. It is advisable to use as many as possible power supply pins for both analog and digital circuits in order to reduce the inductance of bondwire. (Using two bondwires for one pad usually does not reduce the inductance due to the mutual inductance).

In figure 6.20, we show a conceptual pin assignment for a mixed analog-digital chip.

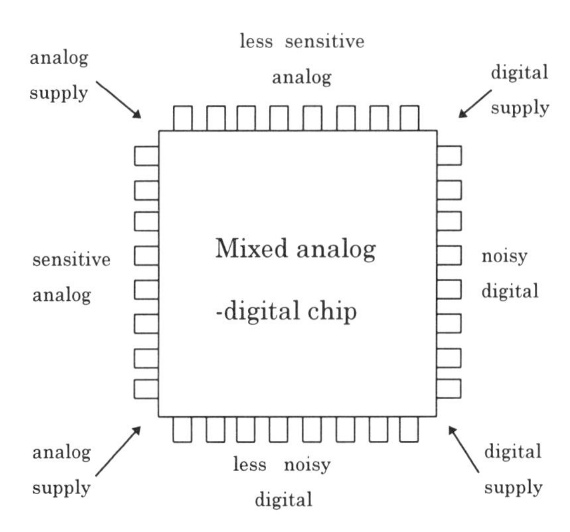

Fig. 6.20. Pin assignment for a mixed analog-digital chip.

For high speed circuits, a small external series resistance can be added to damp the ringing in the supply lines. It is also advisable to consider the PCB layout when we do the pin assignment.

G. Noise coupling in a heavily doped substrate

If the substrate is heavily doped (resistivity ~ 5 mohm-cm), the shielding by using substrate contacts is not as effective as if the substrate is lightly doped (resistivity ~ 10 ohm-cm), while other protection techniques discussed above are still effective. The whole heavily-doped substrate is like a low-ohmic conducting plate, any noise pumped into the substrate can be directly coupled to the sensitive part. It is therefore better to reduce the substrate contacts at the digital blocks to reduce the noise pumped into the substrate. (Since the substrate has a low resistance, we do not need many substrate contacts to avoid the latch-up.) Low-inductance backside contacts can also substantially reduce the noise coupling via the substrate. However, in order to make the backside contacts effective we have to thin the wafer, because the noise wave only travels on the surface and the skin depth is dependent on the frequency [76].

Generally speaking, the noise coupling via the substrate in a heavily-doped substrate is severer and more difficult to control than in a lightly-doped substrate. Fortunately, it is usually possible to choose a lightly-doped substrate for the CMOS process. Substrate doping profiles usually do not change the processing steps significantly. However, it may be easier to trigger the latch-up in a lightly-doped substrate CMOS chip if precaution is not taken.

6.8. SUMMARY

In this chapter, we have discussed practical aspects of SI circuits and systems. Considering the specialties of SI circuits compared to SC circuits, we have discussed the simulation setup for SI circuits, the clocking strategy, the loading effect, and the resetting. The major point is avoiding open-circuiting high impedance outputs of current sources and memory cells. Then we have discussed issues related with the analog layout and especially the matching of analog components. The interdigitized layout style is usually preferred over the common-centroid layout style for SI circuits in that the interdigitized layout is more regular and of less complexity. Sometimes, we can even combine both layout styles, i.e., use the common-centroid layout style locally and interdigitized layout style globally. How to deal with the noise coupling in a mixed analog-digital chip has been another focus in this chapter. Shielding is the common practice to protect the sensitive analog signals. The clock distribution should be done with the consideration of reducing the switching noise. NMOS (or PMOS) switches with dummies are usually preferred over CMOS switches for voltage-sampling switches in that the clock feedthrough in CMOS switches is more sensitive to clock signals. For current-steering switches, the major concern is the signal dependent voltage drop across the current-steering switches. Whether NMOS (or PMOS) or CMOS switches should be used is dependent on the potential change. To avoid the noise coupling via the common supply lines, different supply lines should be used for digital and analog circuits, and supply lines should even be separated for noisy and less noisy digital circuits, for noisy and 'quiet' analog circuits if possible. To have a clean low impedance bias network is another key to high performance analog circuits. The current biasing is usually preferred over the voltage biasing. When we do the pin assignment, it is of great importance to separate noisy digital and sensitive analog pins as far apart as possible. Protection of the sensitive analog pins by grounded neighboring pins is also encouraged.

Chapter VII: Implementation and Measurement of SI Oversampling A/D Converters

7.1. INTRODUCTION

To evaluate the performance of an oversampling A/D converter (or delta-sigma modulator), we need to measure the signal-to-noise-plus-distortion ratio (SNDR), or signal/(noise + THD) within a certain signal band. When we increase the input amplitude, the SNDR increases. If we further increase the input amplitude, the SNDR starts to decrease due to the saturation in the converter. The saturation in the converter introduces strong harmonic distortions and also increases the noise level. The converter will eventually stop working if the input amplitude is too large.

The peak SNDR is the largest SNDR when we sweep the whole input amplitude range. The dynamic range is usually defined as the difference between the largest input amplitude with which the SNDR is acceptable and the smallest input amplitude with which the SNDR is zero. The definition of the largest input amplitude is not unanimous in that large distortions are sometimes acceptable when the input signal is strong for certain applications. For simplicity, we can define the largest input amplitude in an SI oversampling delta-sigma modulator as the full-scale input current. Due to the saturation in the converter, the peak SNDR is always smaller than the dynamic range. With an ideal data converter, the peak SNDR would be equal to the dynamic range.

In this chapter, we will report measurement results of some implemented SI oversampling A/D converters including second-order, two-stage fourth-order, and chopper-stabilized second-order SI delta-sigma modulators. Before we report the measurement results, we will briefly describe the measurement setup. After reporting the measurement results, we will compare the measurement results with theoretical expectations and compare the implemented SI delta-sigma modulators with the state-of-the-art SC delta-sigma modulators.

7.2. MEASUREMENT SETUP

The measurement setups for testing SI oversampling delta-sigma modulators are shown in figure 7.1.

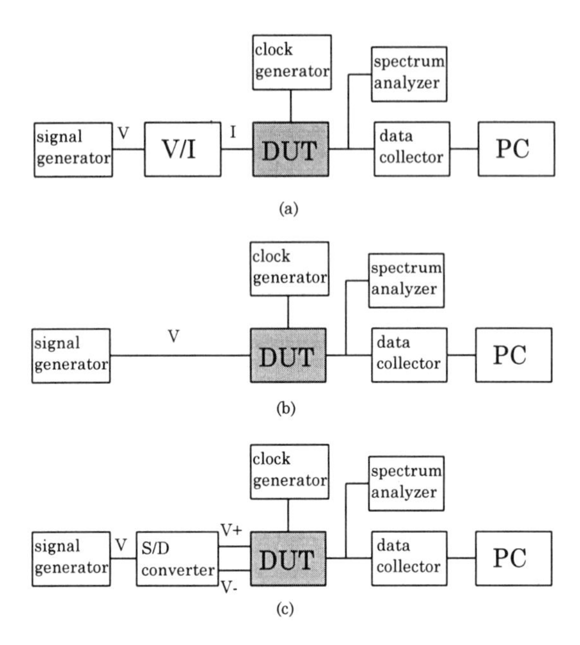

Fig. 7.1. Measurement setups for testing SI oversampling delta-sigma modulators.

The oversampling SI delta-sigma modulator is the device under test (DUT). The digital outputs are collected by a data collector and then processed by a PC. The data collector can be a logic analyzer, or a specialized high speed data buffer. The requirements are the speed and memory depth. Without a decimation filter (or a serial-to-parallel converter) within the DUT or following the DUT, the memory depth of the data collector should be at least 16 K depending on the oversampling ratio. We can then calculate SNR, distortion, and SNDR by using the FFT in the PC or a workstation. It is also a good idea to monitor the digital outputs with a spectrum analyzer, though the spectrum analyzer only gives a qualitative indication of the performance.

If the DUT does not have an on-chip V/I converter, an external V/I converter is needed as shown in figure 7.1 (a). If the DUT have an on-chip V/I converter, the setup shown in figure 7.1 (b) can be used. If the DUT does not have a single-ended to fully-differential (S/D) converter on chip, an external single-ended to fully-differential converter is needed as shown in figure 7.1 (c). If a very clean input signal can not be generated by a signal generator, a low-pass or band-pass filter is needed to follow the signal generator to reduce the harmonics in the signal, otherwise the DUT only measures the distortion and noise from the signal generator.

7.3. SECOND-ORDER SI DELTA-SIGMA MODULATORS

Several second-order SI delta-sigma modulators were implemented according to the architecture of figure 4.3 that has the best settling behavior and the same signal swing in both integrators. The architecture is redrawn in figure 7.2.

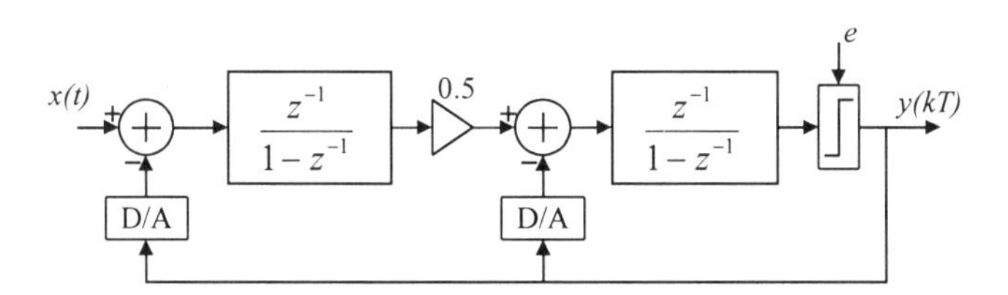

Fig. 7.2. Second-order SI delta-sigma modulator.

Different practical SI memory circuits and other practical building blocks discussed in the previous chapters were used to implement the architecture shown in figure 7.2. The practical aspects discussed in Chapter VI have also been considered in all these implementations.

In table 7.1, we summarize the circuit configurations of six implemented chips.

Table 7.1. Configurations of the second-order SI modulator of figure 7.2.

Name	on-chip V/I	SI circuits	quantizer	D/A conv.
Modulator 1	No	Fig. 3.1	Fig. 5.7 (a)	Fig. 5.9 (b)
Modulator 2	No	Fig. 3.3 (b)	Fig. 5.7 (b)	Fig. 5.9 (a)
Modulator 3	No	Fig. 3.7 + CMFF	Fig. 5.7 (c)	Fig. 5.9 (a)
Modulator 4	Fig. 5.12	Fig. 3.10 + CMFF	Fig. 5.7 (c)	Fig. 5.9 (b)
Modulator 5	No	Fig. 3.16 + CMFF	Fig. 5.7 (d)	Fig. 5.9 (c)
Modulator 6	Fig. 5.12	Fig. 3.21 (b) + CMFF	Fig. 5.7 (c)	Fig. 5.9 (b)

To demonstrate the regularity and modularity of SI circuits, we show in figure 7.3 one photograph of an implemented chip. In the upper part is the 1.2-V second-order SI delta-sigma modulator (No. 5) and in the lower part is a 1.2-V delay line with a single-ended to fully-differential converter at the input and a fully-differential to single-ended converter at the output. The clock generator is in the middle. From this photo it is clear that the regularity and modularity of SI circuits are high.

Fig. 7.3. Photograph of the 1.2-V SI delta-sigma modulator and delay line.

There is no point to go through all the measurements. We only show some measurements and then summarize all the measurement results.

In figure 7.4, we show the measured power spectrum of the 1.2-V SI delta-sigma modulator (No. 5).

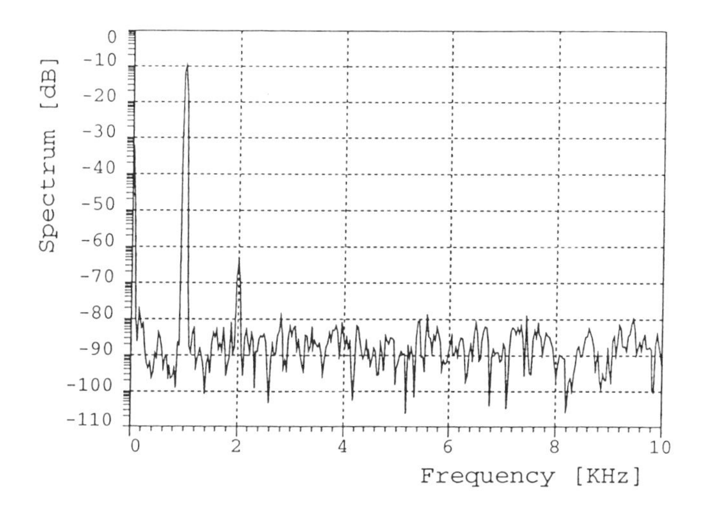

Fig. 7.4. Measured spectrum of the 1.2-V SI modulator (No. 5). SNR = 51 dB and THD = - 55 dBc @ - 10 dBFS.

The clock frequency is 1 MHz and the clock voltages are 0 and 3.3 V. The input is a - 10-dBFS 1-kHz sinusoid. With an oversampling ratio of 64, the measured SNR is 51 dB and the THD is - 55 dBc. The dynamic range is 61 dB, i.e., about 10 bits.

In figure 7.5, we show the measured SNDR vs. the input amplitude of the modulator using the clock feedthrough compensated first-generation SI circuits (No. 1). The input is 2-kHz sinusoid, the clock frequency is 2.45 MHz, and the oversampling ratio is 128. The dynamic range is about 11 bits.

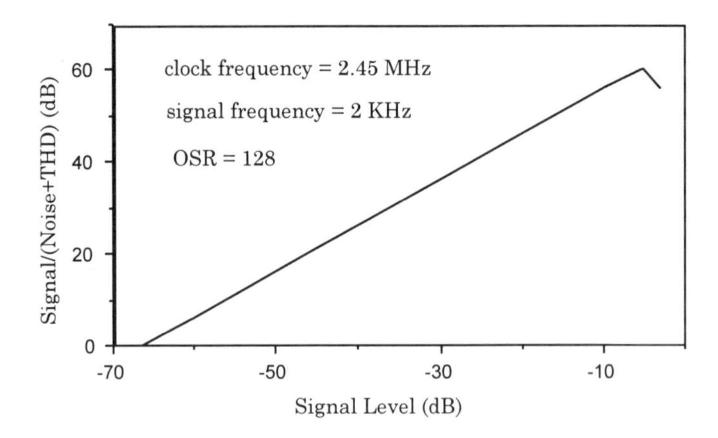

Fig. 7.5. Measured SNDR vs. the input amplitude of the modulator using the clock feedthrough compensated SI memory cells (No. 1).

In Table 7.2, we summarize the measured performances of the six implemented second-order SI delta-sigma modulators of figure 7.2.

Table 7.2. Measured performances of the second-order SI modulators.

Name	Area (mm^2)	V_{dd} (V)	power (mW)	clock (MHz)	OSR	signal BW (kHz)	0-dB level (μA)	DR (bits)
Mod. 1	0.48	3.3	6.6	2.45	128	9.6	20	11
Mod. 2	0.25	3.3	2	2.45	128	9.6	6.5	10
Mod. 3	0.26	3.3	3.2	2.45	128	9.6	6	10.5
Mod. 4	0.5	3.3	20	40	128	156	20	10
Mod. 5	0.47	1.2	0.78	1	64	7.8	9	10
Mod. 6	0.5	3.3	20	40	128	78	20	10

7.4. TWO-STAGE FOURTH-ORDER SI DELTA-SIGMA MODULATORS

Two two-stage fourth-order SI delta-sigma modulators were implemented according to the architecture of figure 4.7. It is redrawn in figure 7.6.

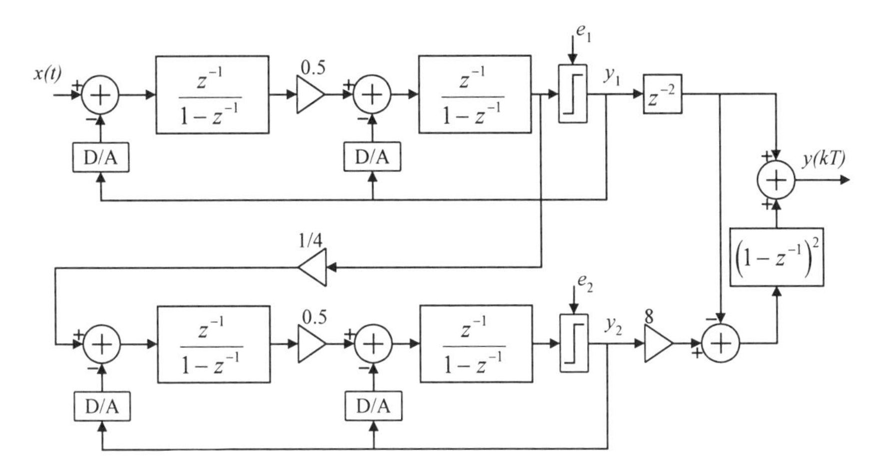

Fig. 7.6. Two-stage fourth-order SI delta-sigma modulator.

The circuit configurations are summarized in table 7.3.

Table 7.3. Configurations of the fourth-order SI modulator of figure 7.6.

Name	on-chip V/I	SI circuits	quantizer	D/A conv.
Modulator 7	Fig. 5.12	Fig. 3.10 + CMFF	Fig. 5.7 (c)	Fig. 5.9 (b)
Modulator 8	Fig. 5.12	Fig. 3.21 (b) + CMFF	Fig. 5.7 (c)	Fig. 5.9 (b)

In figure 7.7, we show the measured power spectrum of modulator 7. The clock frequency is 40 MHz and the input is a - 10-dB 80-kHz sinusoid. The measured SNDR is about 45 dB with an oversampling rate of 32. The dynamic range is about 55 dB.

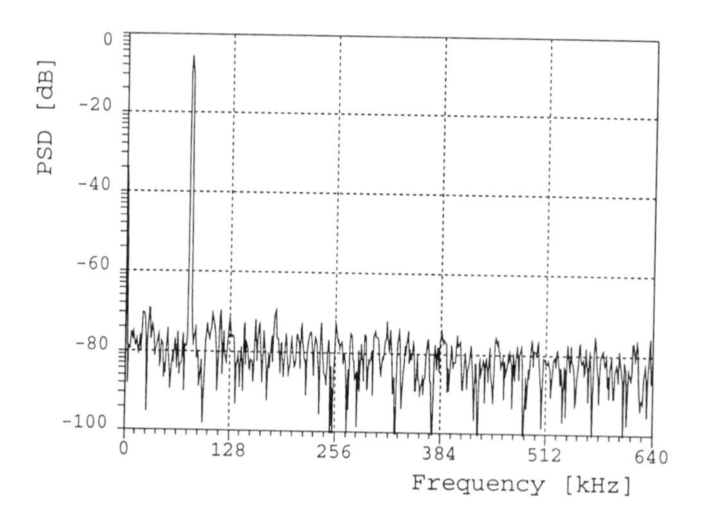

Fig. 7.7. Measured spectrum of the high-speed modulator (No. 7). SNDR = 45 dB @ -10 dBFS with OSR = 32.

In Table 7.4, we summarize the measured performances of the two implemented two-stage fourth-order SI delta-sigma modulators of figure 7.6.

Table 7.4. Measured performances of the fourth-order SI modulators.

Name	Area (mm^2)	V_{dd} (V)	power (mW)	clock (MHz)	OSR	signal BW (kHz)	0-dB level (μA)	DR (bits)
Mod. 7	1.0	3.3	40	40	32	625	20	9
Mod. 8	1.0	3.3	40	40	32	312	20	9

Notice that in modulator 8, the two-step SI circuits are used and therefore the effective sampling rate is 20 MHz, though the master clock frequency is still 40 MHz.

7.5. CHOPPER-STABILIZED SI DELTA-SIGMA MODULATOR

The chopper-stabilized SI delta-sigma modulator of figure 4.15 was also implemented. We redraw the modulator in figure 7.8.

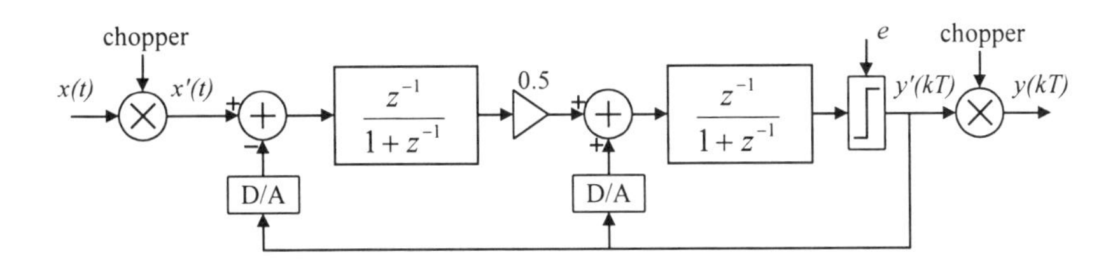

Fig. 7.8. Chopper-stabilized SI delta-sigma modulator.

The circuit configuration is summarized in table 7.5.

Table 7.5. Configuration of the chopper-stabilized SI modulator.

Name	on-chip V/I	SI circuits	quantizer	D/A conv.
Modulator 9	No	Fig. 3.7 + CMFF	Fig. 5.7 (c)	Fig. 5.9 (a)

In figures 7.9 and 7.10, we show the measured output spectra of the chopper-stabilized modulator outputs.

The input signal is a 2-kHz -6-dBFS sinusoid and the clock frequency is 2.45 MHz. Shown in figure 7.9 is the output spectrum before the output chopper stabilization. It is seen in this figure that the signal spectrum is in the band close to half of the sampling frequency. Shown in figure 7.10 is the output spectrum after the output chopper stabilization. It is seen in this figure that the signal spectrum is moved to low frequencies. Notice that the noise at low frequencies is mainly due to the input interface circuit.

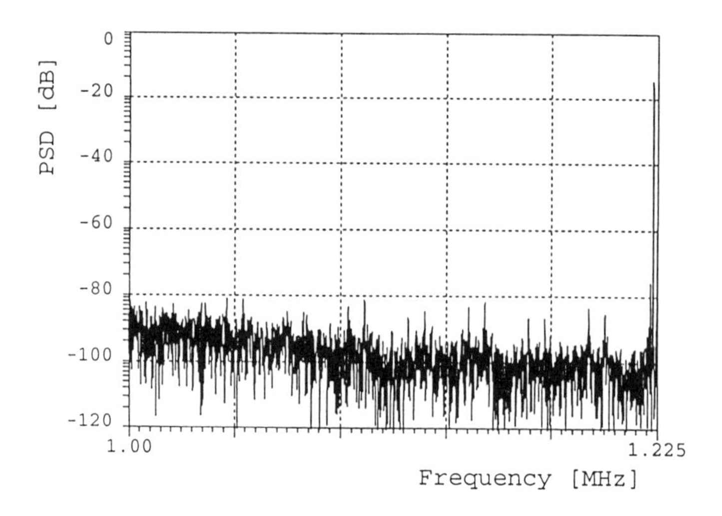

Fig. 7.9. Output spectrum before the output chopper stabilization.

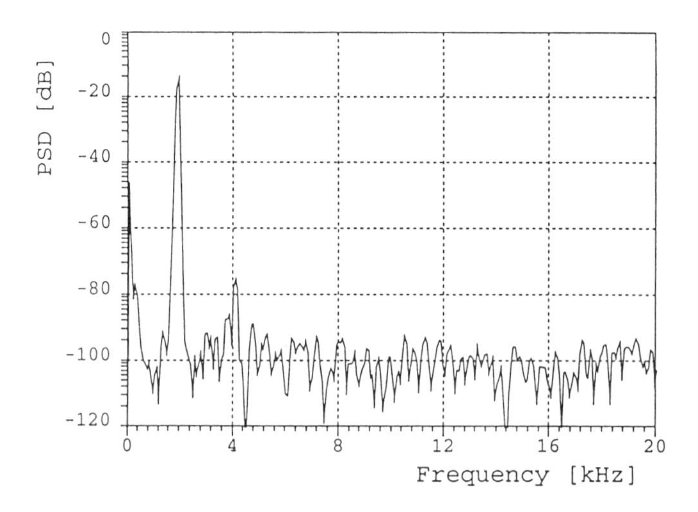

Fig. 7.10. Output spectrum after the output chopper stabilization.

In figure 7.11 we show the measured SNDR vs. the input amplitude of the chopper stabilized SI modulator (No. 9) and the non-chopper stabilized SI modulator (No. 3). The signal is a 2-kHz sinusoid, the clock frequency is 2.45 MHz, and the oversampling ratio is 128. The measured dynamic range for both modulators is about 10.5 bits.

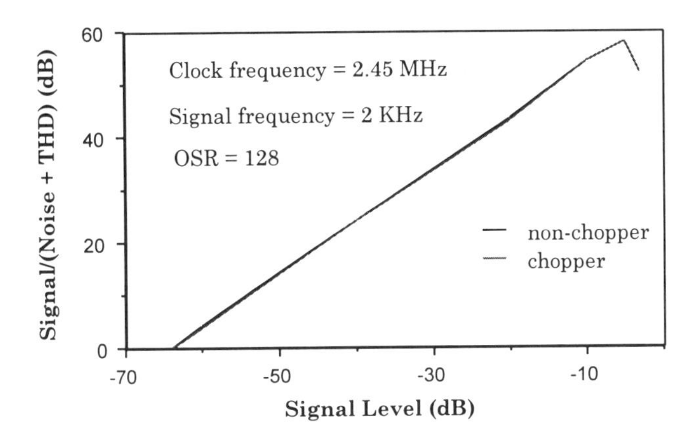

Fig. 7.11. Measured SNDR vs. the input current of the chopper-stabilized and non-chopper-stabilized second-order SI modulators.

In Table 7.6, we summarize the measured performances of the implemented chopper-stabilized SI delta-sigma modulator of figure 7.7.

Table 7.6. Measured performances of the chopper-stabilized modulator.

Name	Area (mm^2)	V_{dd} (V)	power (mW)	clock (MHz)	OSR	signal BW (kHz)	0-dB level (μA)	DR (bits)
Mod. 9	0.26	3.3	3.2	2.45	128	9.6	6	10.5

7.6. COMPARISON WITH THEORETICAL EXPECTATIONS

From the measurement results, we see that the SI delta-sigma modulators achieve a lower dynamic range than what is expected from the system-level simulation. The reason is the thermal noise at the input of the SI modulators.

The thermal noise can be easily calculated by considering the noise generated by the transistors and the noise bandwidth [1]. This has been detailed in Chapter 2.7. Here we just recapitulate the major points.

The total current noise power in an SI memory cell is given by equation (2.53). Therefore, the rms current noise is

$$\overline{i_{rms}} = \sqrt{\frac{\overline{i_n^2}}{\Delta f} \cdot BW_n} = \sqrt{\frac{kT}{C} \cdot \left\{ \frac{2}{3} g_{m0}^2 \cdot \left[1 + \frac{g_{mJ}}{g_{m0}} \right] \right\}} \tag{7.1}$$

where $\frac{\overline{i_n^2}}{\Delta f}$ is the thermal noise power spectral density, BW_n is the noise bandwidth, k is the Boltzmann constant, T is the absolute temperature, C is the total capacitance at the gate of the memory transistor, g_{m0} is the transconductance of the memory transistor, and g_{mJ} is the transconductance of the transistor forming the current source.

The dynamic range is the ratio of the highest input current (full-scale current) and the rms noise. By oversampling, the dynamic range is increased due to the reduced signal bandwidth in respect to the clock frequency. The dynamic range in a thermal-noise limited SI circuit is therefore given by

$$DR = 20 \cdot \log \left(\frac{FS \cdot 1/\sqrt{2}}{\overline{i_{rms}}} \right) + 10 \cdot \log OSR \tag{7.2}$$

where FS is the full-scale input current and OSR is the oversampling ratio. Notice that the coefficient $1/\sqrt{2}$ is due to the calculation of the rms value of the input signal.

The calculated rms current values are in the range of 30 ~ 60 nA for different designs. The difference between the maximum dynamic range governed by the thermal noise and the measured dynamic range is somewhere between 3 ~ 10 dB. Therefore, the thermal noise limits the performance of SI delta-sigma modulators.

An example of the numerical estimations is given below for the 1.2-V SI delta-sigma modulator (No. 5).

The contribution to the noise at the modulator input (the first integrator input) is due to the thermal noise in the SI memory cell and thermal noise in the 1-bit D/A converter fed to the first integrator. The former has a rms value of 15 nA, and the latter has a rms value of 6.5 nA. The total rms noise current present at the modulator input is therefore equal to 16.4 nA. With a peak input current 9 μA, the modulator would achieve a dynamic range of 51 dB within the Nyquist bandwidth. Oversampling by a factor of 64 increases the dynamic range by 18 dB within the signal bandwidth. Therefore, the modulator could achieve a dynamic range of 69 dB. The measured value is about 61 dB. The difference is mainly due to the approximation in estimating the noise power spectral density and the noise bandwidth. It is seen that it is the noise at the input that limits the dynamic range of SI delta-sigma modulators.

The chopper-stabilized SI delta-sigma modulator (No. 9) does not achieve a higher dynamic range than the non chopper-stabilized delta-sigma modulator (No. 3). The reasons are 1) the circuits are second-generation SI circuits and correlated double sampling reduces the low-frequency noise; and 2) the thermal noise determines the noise floor on which the chopper stabilization has no effect. Not being able to demonstrate the advantages of the chopper stabilization at the system level, the chopper-stabilized SI delta-sigma modulator is an interesting alternative to realizing oversampling A/D converters and there is no penalty in complexity except for some chopper switches.

7.7. Comparison with SC Implementations

A. Bandwidth comparison

As evident in the preceding discussions, the thermal noise rather than the quantization noise limits the dynamic range of an SI delta-sigma modulator. However, if we reduce the oversampling ratio (thereby increase the signal bandwidth), the quantization noise will dominate. If the quantization noise rather than the circuit noise dominates, both SI and SC delta-sigma modulators have the same dynamic range given the same architecture and sampling rate.

Shown in figure 7.12 is the simplified noise power spectral density for an SI and SC delta-sigma modulator having the same architecture and sampling rate.

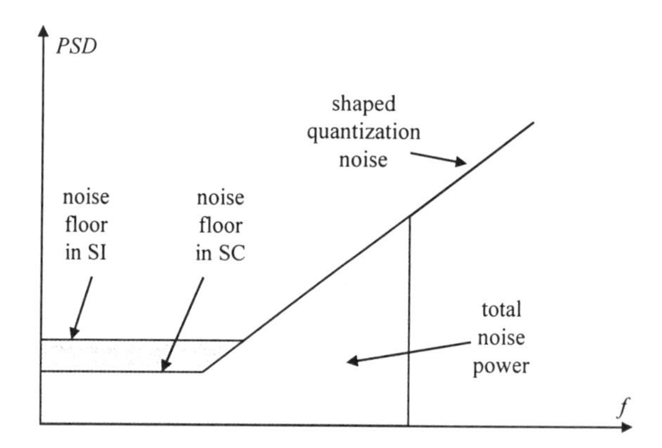

Fig. 7.12. Noise power spectral density in an SI and SC delta-sigma modulator.

Due to the circuit noise, the SI delta-sigma modulator has a higher noise floor at low frequencies. The total noise power within a given bandwidth is the area beneath the power spectral density curve as shown in figure 7.12. If we increase the signal bandwidth (i.e., by reducing the oversampling ratio), the shaped quantization noise will dominate the total noise power. (Notice that the frequency axis is logarithmic.) Therefore, for wide-band applications where the oversampling ratio is usually low and the quantization noise dominates the total noise power, SI delta-sigma modulators are usually advantageous in that they can operate at higher sampling rates than SC counterparts.

Though SC delta-sigma modulators can also operate at a clock frequency up to 50 MHz, they usually require a 5-V supply voltage [77-78]. When the supply voltage is decreased, the speed degrades significantly due to the difficulties in designing high-speed operational amplifiers.

To compare the performance of the state-of-art SI and SC delta-sigma modulators, we have done a search within the INSPECT database. Though there are many publications on SC delta-sigma modulators, we are only interested in the implementations where a lower-than-3.3-V supply is used [79-82]. There are not so many papers on low-voltage SC delta-sigma modulators, and we even include the SC implementation in BiCMOS [79].

Since both the thermal noise and quantization noise may limit the dynamic range of both SI and SC delta-sigma modulators, we make a simplified assumption that the dynamic range is limited by the quantization noise for the reported SC delta-sigma modulators when the oversampling ratio is less than 64 and that the dynamic range is limited by the quantization noise for the reported SI delta-sigma modulators when the oversampling ratio is less than 16. This is a fair assumption in that the circuit noise in an SI delta-sigma modulator is larger than that in an SC delta-sigma modulator, though more accurate assumption should take into consideration the noise shaping order.

When the dynamic range is limited by the quantization noise, the dynamic range decreases by $L + 0.5$ (or L) bits for every doubling of the signal bandwidth for an L-th-order multi-stage (or single-stage) delta-sigma modulator. (Notice that for single-stage architectures, the noise shaping is not as sharp as the multi-stage architectures having the same order noise-shaping loop and therefore we assume that the dynamic range decreases by only L bits for simplicity.) When the dynamic range is limited by the thermal noise, the dynamic range decreases only by 0.5 bit for every doubling of the signal bandwidth independent of the noise shaping order. Based on these, we plot the comparison result in figure 7.13.

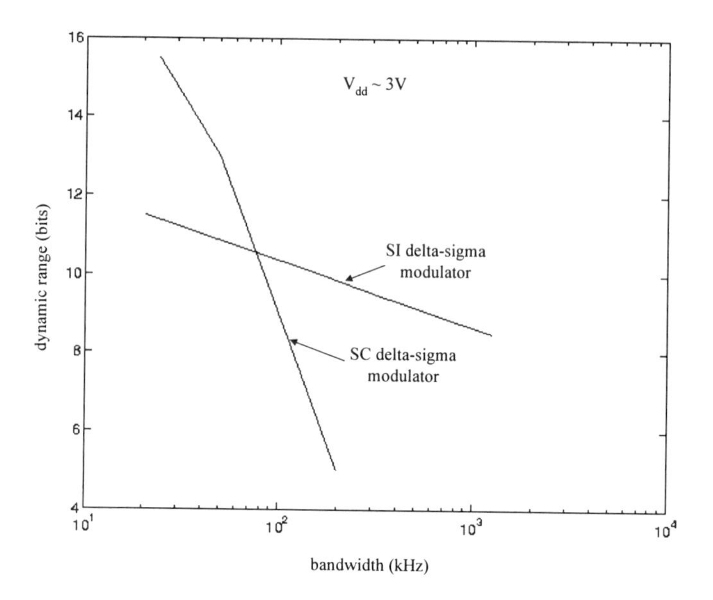

Fig. 7.13. Dynamic range vs. the bandwidth for low-voltage SI and SC modulators.

It is seen that the SI delta-sigma modulators have a higher dynamic range for wide signal bandwidths even compared with the state-of-art SC delta-sigma modulators implemented in BiCMOS, if the supply voltage is reduced below 3.3 V.

B. Supply-voltage comparison

It is possible to operate SC delta-sigma modulators at a very low supply voltage in a low-threshold CMOS process [83]. In a standard CMOS process, SC delta-sigma modulators usually need a supply voltage at least about 2 V due to the difficulties in designing high-performance operational amplifiers with a low supply voltage, while SI delta-sigma modulators can operate at a supply voltage as low as 1.2 V. In figure 7.14, we summarize the requirement of the minimum supply voltage for SI and SC delta-sigma modulators. The signal bandwidth is assumed to be about 8 kHz.

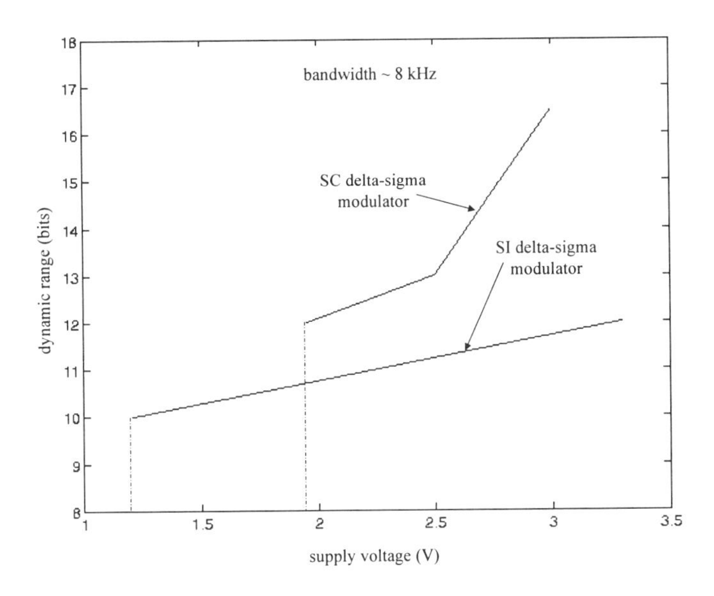

Fig. 7.14. Dynamic range vs. the supply voltage for low-bandwidth SI and SC delta-sigma modulators.

It is seen in figure 7.14 that SC delta-sigma modulators stop working at a supply voltage around 2 V while SI delta-sigma modulators can still deliver a 10-bit dynamic range at a supply voltage of 1.2 V, assuming that no low-threshold devices are available. SI delta-sigma modulators can operate at a much lower supply voltage thanks to the absence of operational amplifiers. Due to the use of operational amplifiers, it is very difficult for SC delta-sigma modulators to operate at a supply voltage lower than twice the threshold voltage.

7.8. SUMMARY

In this chapter, after describing the measurement setup, we have presented measurement results of 9 different SI delta-sigma modulators including high-sampling-rate (40 MHz @ 3.3 V) and ultra-low-voltage (1.2 V) SI delta-sigma modulators. We have also compared the measurement results with theoretical expectations, confirming that the noise at the modulator input limits the dynamic range of SI delta-sigma modulators. Compared with the state-of-the-art SC delta-sigma modulators, SI delta-sigma modulators usually have a much wider signal bandwidth at a low supply voltage (~3.3 V) and the supply voltage for SI modulators can be reduced much further without degrading the performance. Therefore, we have confirmed that the SI technique is suitable for high speed and/or low-voltage operations.

Chapter VIII: Conclusions

In this book, we have discussed the SI technique and its application to the design of oversampling A/D converters. We have presented many low-voltage and/or high-speed design examples and measured circuits.

The book has first introduced the principle of the SI technique and discussed the nonidealities of the SI technique. Then practical SI circuits with measurement results have been presented. To design SI oversampling A/D converters, system design issues have been addressed followed by discussions of building blocks. Before we present measurement results of SI oversampling A/D converters, we have also covered various practical aspects of SI circuits and systems.

As discussed in the book, the SI technique is a viable alternative to the traditional SC technique with both pros and cons. The major advantages of the SI technique are the high-speed and low-supply-voltage operations, while the major disadvantages are the relatively larger noise and less accuracy compared with the SC technique. In oversampling A/D converters, the advantages of the SI technique can be fully utilized while the disadvantages can be suppressed to a certain extent.

To demonstrate that the SI technique is a viable circuit design technique, six test chips of SI memory cells and delay lines were implemented and measured based on the practical SI circuits discussed in this book. For audio applications, the designed first-generation SI memory cell with clock feedthrough compensated has a THD as low as -65 dBc. The high-speed SI circuits can deliver a THD of -47 dBc at a 100-MHz clock frequency with a single 3.3-V supply voltage. For ultra low-voltage applications, the designed SI circuits can deliver a THD of - 48 dBc with a supply voltage of only 1.2 V.

By tailoring delta-sigma modulator architectures for the SI implementation, by using the practical SI circuits, circuit techniques, and other building blocks, and by considering practical aspects of SI circuits and systems, we have designed, implemented and measured nine SI oversampling delta-sigma modulators including high-sampling-rate (40 MHz @ 3.3 V) and ultra-low-voltage (1.2 V) SI delta-sigma modulators. The typical dynamic range is usually 9 ~ 12 bits depending on the bandwidth. We have also done a comparison of the state-of-the-art SI and SC delta-sigma modulators by searching the INSPECT database. Though we do not claim that the comparison is exhaustive, it is evident from this comparison that SI delta-sigma modulators usually have a much wider signal bandwidth and higher dynamic range at a low supply voltage (~3.3 V) and the supply voltage for SI modulators can be reduced much further (less than twice the threshold voltage) without degrading the performance. Therefore, we have confirmed that the SI technique is suitable for high speed and/or low-voltage operations.

References

[1] C. Toumazou, J. B. Hughes, and N. C. Battersby, *Switched-currents: an analogue technique for digital technology.* Peter Peregrinus Ltd., 1993.

[2] S. J. Daubert, D. Vallancourt, and Y. P. Tsividis, "Current copier cells," *Electronics Letters*, Vol. 24, No. 25, pp. 1560-1562, Dec. 1988.

[3] J. B. Hughes, N. C. Bird, and I. C. Macbeth, "Switched Currents - a new technique for analog sampled-data signal processing," in proc. *IEEE International Symposium on Circuits and Systems*, Portland, Oregon, pp. 1584-1587, May. 1989.

[4] P. E. Allen and D. R. Holberg, *CMOS analog circuit design*, Holt, Rinehart and Winston, Inc., 1987.

[5] R. Unbehauen and A. Cichocki, *MOS switched-capacitor and continuous-time integrated circuits and systems*, Springer-Verlag, 1989.

[6] A. T. Behr, M. C. Schneider, S. N. Filho, and C. G. Montoro, "Harmonic distortion caused by capacitors implemented with MOSFET gates," *IEEE Journal of Solid-State Circuits*, vol. SC-27, pp. 1470-1475, 1992.

[7] J. C. M. Bermudez, M. C. Schneider, and C. G. Montoro, "Linearity of switched-capacitor filters employing nonlinear capacitors," in proc. *IEEE International Symposium on Circuits and Systems*, pp. 1211-1214, May. 1992.

[8] H. Yoshizawa and G. C. Temes, "High-linearity switched-capacitor circuits in digital CMOS technology," in proc. *IEEE International Symposium on Circuits and Systems*, pp. 1029-1032, May. 1995.

[9] J. L. McCreary, "Matching properties, and voltage and temperature dependence of MOS capacitors," *IEEE J. Solid-State Circuits*, vol. SC-16, pp. 608-616, 1981.

[10] N. Tan, "Fourth-order SI delta-sigma modulators for high-frequency applications," *IEE Electronics Letters*, Vol. 31, No. 5, pp. 333-334, Mar. 1995.

[11] N. Tan, "A 1.2-V 0.8-mW SI ΔΣ A/D converter in standard digital CMOS process," In *Proc. 21st European Solid-State Circuits Conference (ESSCIRC'95)*, Lille, France, pp. 150-153, Sept. 1995.

[12] N. Tan, G. Amozandeh, A. Olson, and H. Stenström, "Current scaling technique for high dynamic range switched-current delta-sigma modulators," *IEE Electronics Letters*, Vol. 32, No. 15, pp. 1331-1332, July 1996.

[13] M. Gustavsson and N. Tan, "A new pipeline A/D converter architecture and its low-voltage implementation for DECT," in proc. *IEEE Norchip'96 conference*, pp. 114-121, Nov. 1996. Helsinki, Finland.

[14] I. H. H. Jorgensen and G. Bogason, "A 3rd order low power switched current ΣΔ-modulator," in proc. *IEEE Norchip'96 conference*, pp. 89-96, Nov. 1996. Helsinki, Finland.

[15] G. E. Saether, C. Toumazou, G. Taylor, K. Eckersall, and I. M. Bell, "Built-in self test of S2I switched current circuits," *International Journal of Analog Integrated Circuits and Signal Processing*, pp. 25-30, Jan. 1996.

[16] N. Tan, "Switched-current delta-sigma A/D converters," *International Journal of Analog Integrated Circuits and Signal Processing*, pp. 7-24, Jan. 1996.

[17] N. Tan, *Oversampling A/D converters and current-mode techniques*, Ph. D dissertation, Department of Electrical Engineering, Linköping University, Linköping, Sweden, 1994.

[18] B. Kamath, R. Meyer, and P. Gray, "Relationship between frequency response and settling time of operational amplifier," *IEEE J. Solid-State Circuits*, vol. SC-9, pp. 347-352, Dec. 1974.

[19] B. Jonsson, *Applications of the Switched-Current Technique*, Tekn. Lic. dissertation, Department of Electrical Engineering, Linköping University, Linköping, Sweden 1994.

[20] P. J. Crawley and G. W. Roberts, “Predicting harmonic distortion in switched-current memory circuits,” *IEEE Trans. Circuits and Syst.*, Vol. 41, pp. 73-86, Feb. 1994.

[21] H. Träff and S. Eriksson, ”On the clocking of discrete time circuits,” *Report LiTH-ISY-R-1590*, Linköping University, Sweden, March 1993.

[22] P. R. Gray and R. G. Meyer, *Analysis and Design of Analog Integrated Circuits*, Third edition, John Wilet & Sons, Inc., 1993.

[23] C. Eichenberger and W. Guggenbuhl, ”On charge injection in analog MOS switches and dummy compensation technique,” *IEEE Transactions on Circuits and Systems*, vol. 37, No. 2, pp. 256-264, Feb., 1990.

[24] H. C. Yang, T. S. Tiez, and D. J. Allstot, ”Current-feedthrough effects and cancellation techniques in switched-current circuits,” in *Proc. IEEE International Symposium on Circuits and Systems*, pp. 3186-3188, May 1990.

[25] B. Jonsson and S. Eriksson, “A new clock-feedthrough compensation scheme for switched-current circuits,” *Electron. Lett.,* Vol. 29, pp. 1446-47, Aug. 1993.

[26] B. Jonsson and S. Eriksson, “A low-voltage wave SI filter implementation using improved delay elements,” in *Proc. IEEE International Symposium on Circuits and Systems*, Vol. 5, pp. 305-308, May 1994.

[27] N. Tan, B. Jonsson, and S. Eriksson, “3.3-V 11-bit delta-sigma modulator using first-generation SI circuits,” *Electron. Lett.*, pp. 1819-1821, Oct. 1994.

[28] J. B. Hughes and K. W. Moulding, “Switched-current signal processing for video frequencies and beyond,” *IEEE J. Solid-State Circuits*, vol. 28, pp. 314-322, Mar. 1993.

[29] N. Tan and S. Eriksson, “A fully differential switched-current delta-sigma modulator using a single 3.3-V power-supply voltage,” in *Proc.*

IEEE International Symposium on Circuits and Systems, Vol. 5, pp. 485-588, May 1994.

[30] N. Tan and S. Eriksson, "A low-voltage switched-current delta-sigma modulator," *IEEE J. Solid-State Circuits*, vol. 30, pp. 599-603, May 1995.

[31] P. M. Sinn and G. W. Roberts, "A comparison of first and second generation switched-current cells," in *Proc. IEEE International Symposium on Circuits and Systems*, Vol. 5, pp. 301-304, May, 1994.

[32] N. Tan and S. Eriksson, "Low-voltage fully differential class-AB SI circuits with common-mode feedforward," *Electron. Lett.*, pp. 2090-2091, Dec. 1994

[33] N. Tan and S. Eriksson, "Low-voltage low-power switched-current circuits and systems," in *Proc. European Design and Test Conference*, pp. 100-104, March, 1995.

[34] N. Tan, "3.3-V class-AB switched-current circuits and systems," *IEE Proceedings, Part G, Circuits Devices Syst.*, Vol. 143, No. 2, pp. 97-102, April 1996.

[35] N. Battersby and C. Toumazou, "Class AB switched-current memory for analogue sampled data systems," *Electron. Lett.*, Vol. 27, pp. 873-875, May. 1991.

[36] H. Träff and S. Eriksson, "Class A and AB compact switched-current memory circuits," *Electron. Lett.*, Vol. 29, pp. 1446-47, Aug. 1993.

[37] J. B. Hughes and K. W. Moulding, "S^2I: a switched-current technique for high performance," *Electron. Lett.*, vol. 29, pp. 1400-1401, Aug. 1993.

[38] C. Toumazou and S. Xiao, "n-step charge injection cancellation scheme for very accurate switched-current circuits, " *Electron. Lett.*, vol. 30, pp. 680-681, Apr. 1994.

[39] P. Shah and C. Toumazou, "A new BiCMOS technique for very fast discrete-time signal processing," in Proc. *1995 International Symposium on Circuits and Systems*, pp. 323-326.

[40] W. Guggenbuhl, J. Di, and J. Goette, "Switched-current memory circuits for high precision applications," *IEEE J. Solid-State Circuits*, vol. SC-29, pp. 1108-1116, Sept., 1994.

[41] J. C. Candy and G. C. Temes, *Oversampling Delta-Sigma Data Converters: Theory, Design and Simulation.* IEEE Press, 1992.

[42] S. J. Daubert and D. Vallancourt, "A transistor-only current-mode sigma-delta modulator," *IEEE J. Solid-State Circuits*, vol. 27, pp. 821-830, May 1992.

[43] P. J. Crawley and G. W. Roberts, "Switched-current sigma-delta modulation for A/D conversion," in *IEEE Proc. ISCAS'92*, pp. 1320-1323, May 1992.

[44] C. Toumazou and G. Saether, "Switched-current circuits and systems," in *Proc. IEEE International Symposium on Circuits and Systems*, Vol. Tutorials, pp. 459-486, May 1994.

[45] N. Tan and S. Eriksson, "A fully differential switched-current delta-sigma modulator using a single 3.3-V power-supply voltage," in *Proc. IEEE International Symposium on Circuits and Systems*, Vol. 5, pp. 485-588, May 1994.

[46] V. F. Dias, V. Liberali, and F. Maloberti, *TOSCA user's guide*, version 0.9 (e), Dec. 1992.

[47] T. Karema, *Oversampling A/D and D/A converters using one-bit sigma-delta modulation techniques*, Ph.D. Dissertation, Tampere University, Finland, Aug. 1994.

[48] H. Baher and E. Affif, "Novel fourth-order sigma-delta converter," *Electron. Lett.*, vol. 28, pp. 1437-8, July 1992.

[49] F. Medeiro, B. Perez-Verdu, A. Rodriguez-Vazquez, and J. L. Huertas, "Design consideration for a fourth-order switched-capacitor sigma-delta modulator," in *Proc. 11th European Conf. Circuit Theory and Design*, pp. 1607—12, Sept. 1993.

[50] T. Karema, T. Ritoniemi, and H. Tenhunen, "An oversampling sigma-delta A/D converter circuit using two-stage fourth-order modulator," in

Proc. IEEE International Symposium on Circuits and Systems, pp. 3279-82, May 1990.

[51] N. Tan, "Oversampling delta-sigma modulation for A/D converters and its new circuit realizations," *Licentiate Thesis*, Linköping University, Mar. 1993.

[52] N. Tan and S. Eriksson, "New modeling method for 1-bit delta-sigma modulators," in *Proc. 11th European Conf. Circuit Theory and Design*, Davos, Switzerland, pp. 1361-1366, Sept. 1993.

[53] H. Tenhunen, "Analog and digital circuit design for high-accuracy converters," *course notes*, Lausanne, Switzerland, July, 1991.

[54] Y.-H. Chang, C.-Y. Wu, and T.-C. Yu, "Chopper-stabilized sigma-delta modulator," in *Proc. IEEE International Symposium on Circuits and Systems*, pp. 1286-1289, May 1993.

[55] N. Tan, G. Amozandeh, A. Olson, and H. Stenström, "Current scaling technique for high dynamic range switched-current delta-sigma modulators," *IEE Electronics Letters*, Vol. 32, No. 15, pp. 1331-1332, July 1996.

[56] H. Träff, "Novel approach to high speed CMOS current comparators," *Electron. Lett.*, vol. 28, pp. 310-311, Jan. 1992.

[57] R. Gregorian and G. Temes, *Analog MOS Integrated Circuits for Signal Processing*, John Wiley & Sons, Inc., 1986.

[58] N. C. Battersby, *Switched-current techniques for analogue sampled-data signal processing*, Ph.D dissertation, Imperial College, London, U.K., 1993.

[59] Current Conveyor Amplifier, data sheet, LTP Electronics, U.K.

[60] S. T. Dupine and M. Ismail, "High frequency CMOS transconductors," in the book, *Analog IC Design: the Current-Mode Approach*, (Eds: C. Toumazou, F. J. Lidgey, and D. G. Haigh), Peter Peregrinus, Ltd, 1990.

[61] Y. P. Tsividis, "Integrated continuos-time filter design - an overview," *IEEE J. Solid-State Circuits*, vol. 29, pp. 166-176, March 1994.

[62] M. Gustavsson and N. Tan, "A complete analog interface for current-mode A/D converters in a digital CMOS process," in Proc. *IEEE NORSIG'96,* Helsinki, Finland, Sept. 1996.

[63] H. Meleis and P. L. Fur, "A novel architecture design for VLSI implementation of an FIR decimation filter," in *IEEE Proc. ISCAS'85*, pp. 1380-1383, March 1985.

[64] S. Chu and C. S. Burrus, "Multirate filter designs using comb filters," *IEEE Trans. Circuits and Sys.*, vol. CAS-31, pp. 913-924, Nov. 1984.

[65] E. Dijkstra, O. Nys, C. Piguet, and M. Degrauwe, "On the use of module arithmetic comb filters in sigma delta modulators," in *IEEE Proc. ISCAS'88*, pp. 2001-2004, April 1988.

[66] T. Saramäki and H. Tenhunen, "Efficient VLSI-realizable decimators for sigma-delta analog-to-digital converters," in *IEEE Proc. ISCAS'88*, pp. 1525-1528, June 1988.

[67] R. E. Crochiere and L. R. Rabiner, *Multirate Digital Signal Processing.* Prentice-Hall, Inc., 1983.

[68] Y. Gao and N. Tan, "Decimation comb filter for oversampling A/D converters," *internal Ericsson report*, Nov. 1996.

[69] L. Wanhammar, *DSP Integrated Circuits.* Pre-publication edition, 1993.

[70] N. Tan, S. Eriksson, and L. Wanhammar, "A novel bit-serial design of comb filters for oversampling A/D converters," In *Proc. IEEE International Symposium on Circuits and Systems (ISCAS'94)*, London, UK, Vol. 4, pp. 259-262, May, 1994.

[71] Y. Nakagome, et. al., "Experimental 1.5-V 64-Mb DRAM," *IEEE J. Solid State Circuits*, vol. 26, April 1991, pp. 465-472.

[72] N. Tan, "Design and Implementation of Digital-to-Analog Converters for Telecommunication," *ERICSSON Report, KI/EKA/MERC-96:078*, 1996-12-05.

[73] N. Tan, "A 1.5-V 3-mW 10-b 50-Ms/s CMOS DAC with low distortion and low intermodulation in standard digital CMOS process," in proc. *IEEE Custom Integrated Circuit Conference (CICC)*, California, USA, May, 1997.

[74] G. Amozandeh, A. Olson, and H. Stenström, "SCARLET: a switched current sigma delta modulator," *ERICSSON Report, KI/EKA/1/0362-PCP 202 42 Uen*, 1996-01-25.

[75] S. M. Sze, *VLSI Technology*, McGraw Hill, New York, 1983.

[76] N. K. Verghess, T. J. Schmerbeck, D. J. Allstot, *Simulation Techniques and Solutions for Mixed-Signal Coupling in Integrated Circuits*, Kluwer Academic Publishers, Co., Norwell, MA, 1994.

[77] B. P. Brandt and B. A. Wooley, "A 50-Mhz multibit sigma-delta modulator for 12-b 2-Mhz A/D conversion," *IEEE J. Solid-State Circuits*, vol. 26, no. 12, Dec. 1991, pp. 1746-56.

[78] S. Ingalsuo, T. Ritoniemi, T. Karema, and H. Tenhunen, "A 50 MHz cascaded sigma-delta A/D modulator," in *Proc. IEEE 1992 Custom Integrated Circuits Conference*, New York, USA, 1992, pp. 16.3/1-4.

[79] T. Ritoniemi, E. Pajarre, S. Ingalsuo, T. Husu, V. Eerola, and T. Saramiki, "A stereo audio sigma-delta A/D converter," *IEEE J. Solid-State Circuits*, vol. 29, no. 12, Dec. 1994, pp. 1514-23.

[80] S. Kiriaki, "A 0.25-mW sigma-delta modulator for voice-band applications," in *Proc. 1995 IEEE Symposium on VLSI circuits*, Tokyo, Japan, pp. 35-6.

[81] S. Au and B. Leung, "A 1.95-V 0.34-mW 12-b sigma-delta modulator stabilized by local feedback loops," *IEEE J. Solid-State Circuits*, vol. 32, no. 3, Mar. 1997, pp. 321-8.

[82] J. Sauerbrey and M. Mauthe, "12 bit 1 mW delta-sigma modulator for 2.4 V supply voltage," In *Proc. 21st European Solid-State Circuits Conference (ESSCIRC'95)*, Lille, France, pp. 138-41, Sept. 1995.

[83] Y. Matsuya and J. Yamada, "1 V power supply low-power consumption A/D conversion technique with swing-suppression noise shaping," *IEEE J. Solid-State Circuits*, vol. 29, no. 12, Dec. 1994, pp. 1524-30.

Index

—H—

—I—

—L—

—M—

—N—

—O—

—P—

—R—

—S—

—T—

—U—